Angelfish Research

Angelfish Research

TRANSLATING FINDINGS FOR HUMAN HEALTH

Olivia K.

Mohammed Altaf Hussain

Contents

INDEX   1

INTRODUCTION   3

Chapter 1   13

Chapter 2   27

Chapter 3   41

Chapter 4   56

Chapter 5   73

Chapter 6   86

Chapter 7   99

Chapter 8   112

INDEX

Introduction

1. A brief overview of the significance of angelfish research in the biomedical field.
2. Establishing the purpose of the book: translating angelfish findings to benefit human health.

Chapter 1 The Angelfish Model

1.1 Overview of angelfish species used in biomedical research.
1.2 Unique characteristics that make angelfish a valuable model organism.
1.3 Historical context and evolution of angelfish research.

Chapter 2 Genetics and Genomic Insights

2.1 Exploration of angelfish genetics and its parallels to human genetics.
2.2 Comparative analysis of genomic findings between angelfish and humans.
2.3 Potential applications in understanding genetic disorders.

Chapter 3 Cardiovascular Discoveries

3.1 Examination of angelfish cardiovascular systems and similarities to human hearts.
3.2 Insights into heart development and potential implications for cardiac health.
3.3 Implications for cardiovascular disease research.

Chapter 4 Neurological Revelations

4.1 Unraveling the neurobiology of angelfish and its relevance to human neuroscience.
4.2 Studying neurodevelopment and insights into neurological disorders.
4.3 Applications in drug discovery for neurological conditions.

Chapter 5 Drug Discovery and Therapeutics

5.1Showcase of angelfish as a model for drug development.

5.2Translating findings into potential therapeutic interventions.

Chapter 6 Immunology and Inflammatory Responses

6.1Analyzing the immune system of angelfish and its connection to human immunity.

6.2Insights into inflammatory processes and autoimmune disorders.

6.3Potential advancements in immunotherapy.

Chapter 7 Metabolic Pathways and Diseases

7.1Investigating metabolic pathways in angelfish and their relevance to human metabolism.

7.2Implications for understanding and treating metabolic disorders.

7.3Dietary and lifestyle recommendations based on angelfish research.

Chapter 8 Translational Challenges and Future Directions

8.1Addressing challenges in translating angelfish findings to human health.

8.2Proposing future directions for research and collaboration.

8.3Ethical considerations and implications for medical practice.

INTRODUCTION

In the tremendous scene of logical request, model creatures have demonstrated to be priceless torchbearers, enlightening the complexities of life's central cycles. One such illuminating presence in the domain of biomedical examination is the angelfish (Pterophyllum spp.), an animal varieties that has risen above its sea-going environment to turn into a promising model for grasping hereditary qualities, physiology, and, strikingly, its translational potential for human wellbeing. This story sets out on an excursion through the flows of angelfish research, digging into its verifiable roots, investigating its novel qualities, and exploring the genomic waters that interface this rich species to significant ramifications for human prosperity.

1.1 The Verifiable Embroidery of Angelfish Exploration

The account of angelfish research finds its foundations woven into the texture of logical interest. While angelfish might have at first enamored aquarium fans with their effortless blades and lively varieties, their importance as a model organic entity unfurled in the labs of scientists trying to reveal the mysteries of life at a sub-atomic level. The verifiable direction of angelfish research uncovers a slow acknowledgment of its true capacity as in excess of an enthralling aquarium occupant yet as a living window into the complexities of hereditary and physiological cycles.

Early examination on angelfish essentially revolved around figuring out its way of behaving, environment, and fundamental science. As researchers dug into the exceptional qualities and ways of behaving of various angelfish species, a more profound appreciation for the expected utilizations of this creature in research started to arise. The change from observational examinations to controlled research facility tests denoted a significant second, opening the way to precise examinations concerning the hereditary and sub-atomic underpinnings of angelfish physiology.

1.2 The One of a kind Qualities of Angelfish: A Model Living being Divulged

What separates angelfish as a model organic entity, and why has it turned into a point of convergence for logical request? The response lies in the mixture of attributes that make angelfish a remarkable subject for examination.

1.2.1 Hereditary Pliability:

One of the foundations of a model organic entity is its hereditary pliability, permitting specialists to control and concentrate on unambiguous qualities effortlessly. Angelfish, with its clear cut hereditary qualities and moderately short age time, gives a worthwhile stage to investigating quality capability and guideline.

The approach of cutting edge hereditary control strategies, like CRISPR/Cas9, has additionally upgraded the accuracy with which specialists can adjust the angelfish genome, opening roads for designated examinations.

1.2.2 Preservation of Organic Cycles:

While angelfish and people might appear completely different, a more critical look uncovers an astonishing protection of essential natural cycles. From metabolic pathways to hereditary flagging, angelfish share shared view with people, offering scientists a similar focal point through which to concentrate on these cycles. This protection considers the extrapolation of discoveries from angelfish studies to possible applications in human wellbeing.

1.2.3 Ecological Awareness:

The responsiveness of angelfish to ecological circumstances adds a layer of intricacy to explore that reflects the complexities of genuine situations. Angelfish answer changes in water quality, temperature, and living space structure, offering analysts the chance to concentrate on the effect of ecological variables on physiological cycles. This awareness broadens the translational capability of angelfish research, as discoveries might reveal insight into how natural elements impact human wellbeing and infection.

1.2.4 Remarkable Physiological Elements:

Past its hereditary and ecological properties, angelfish have extraordinary physiological highlights that make them especially charming for research. From their cardiovascular framework to neurological capabilities, angelfish display variations that line up with the physiological difficulties of their amphibian climate. Disentangling these variations extends how we might interpret angelfish science as well as gives experiences into equal physiological cycles in people.

1.3 Exploring the Genomic Waters: From Angelfish to Human Wellbeing

As the period of genomics unfolded, the angelfish wound up at the front of examinations concerning the sub-atomic underpinnings of life. The sequencing of the angelfish genome denoted a vital second, opening a gold mine of hereditary data that specialists could now investigate. This genomic guide prepared for a more profound comprehension of the hereditary design of angelfish and, basically, its pertinence to human genomics.

1.3.1 Angelfish Genomics: An Outline for Revelation:

The sequencing of the angelfish genome proclaimed another period in angelfish research, furnishing specialists with an exhaustive outline of its hereditary cosmetics.

Genomic studies have since uncovered qualities related with essential organic cycles, revealing insight into the atomic apparatus that administers advancement, proliferation, and reaction to natural improvements. The abundance of genomic information has extended how we might interpret angelfish science as well as situated this species as a hereditary partner in the journey for disentangling the secrets of human hereditary qualities.

1.3.2 Equals to Human Genomics: Connecting the Species Separation:

The disclosure of hereditary equals among angelfish and people has arisen as a weighty part of angelfish research. Shared hereditary components, saved pathways, and likenesses in quality articulation designs have become central issues of convergence, permitting scientists to draw significant examinations between the two species. These equals hold huge translational potential, offering a scaffold across the animal varieties gap and making ready for applications in human wellbeing.

1.4 The Translational Odyssey: From Angelfish Discoveries to Human Wellbeing

A definitive commitment of angelfish research lies in its capacity to rise above the limits of the lab and make an interpretation of discoveries into significant applications for human wellbeing. The translational excursion is an intricate odyssey that includes exploring logical, moral, and down to earth contemplations to overcome any barrier between angelfish science and human medication.

1.4.1 Illness Displaying and Helpful Experiences:

Angelfish, with its hereditary pliability and rationed natural cycles, fills in as an important model for sickness research. By presenting hereditary alterations that imitate human circumstances, scientists can make angelfish models of explicit infections. These models give a stage to concentrating on sickness systems, testing likely restorative mediations, and acquiring experiences into the hereditary premise of human problems.

1.4.2 Cardiovascular Disclosures: Unwinding the Main issue at hand:

The cardiovascular framework, a key part of both angelfish and human physiology, has turned into a point of convergence of translational examination. Studies investigating the complexities of angelfish cardiovascular turn of events and capability offer bits of knowledge that stretch out past sea-going domains. Examinations among angelfish and human hearts give a one of a kind vantage highlight grasping cardiovascular illnesses and concocting novel restorative procedures.

1.4.3 Neurological Disclosures: Disentangling the Angelfish Mind and Then some:

The neurobiology of angelfish has arisen as a rich territory for investigation, with suggestions venturing into the domains of human neuroscience.

Disentangling the complexities of angelfish neurodevelopment, conduct, and mental capabilities gives an establishment to figuring out equal cycles in the human cerebrum. Bits of knowledge gathered from angelfish reads up hold guarantee for propelling comprehension we might interpret neurological issues and the advancement of designated intercessions.

1.4.4 Medication Disclosure and Therapeutics: From Tank to Facility:

The expected utilizations of angelfish research reach out into the area of medication disclosure and restorative mediations. The ID of bioactive mixtures, investigation of novel medication targets, and testing of drug specialists in angelfish models add to the advancement of expected treatments for human illnesses. Angelfish, in this unique situation, fills in as a powerful proving ground for making an interpretation of revelations into unmistakable clinical arrangements.

1.4.5 Immunology and Provocative Reactions: Monitoring the Equilibrium:

Angelfish, with their mind boggling invulnerable frameworks, offer a stage for concentrating on immunology and incendiary reactions. Experiences acquired from angelfish studies contribute not exclusively to grasping fundamental insusceptible systems yet in addition to disentangling the intricacies of immune system problems. The possible progressions in immunotherapy motivated by angelfish research hold guarantee for changing the scene of human safe related sicknesses.

1.4.6 Metabolic Pathways and Illnesses: Exploring the Phone Territory:

The investigation of metabolic pathways in angelfish gives a focal point through which specialists can dig into the cell territory of energy guideline. Examinations concerning metabolic cycles, supplement use, and variations to ecological changes offer experiences into human metabolic issues. Angelfish, as a model creature, adds to the more extensive comprehension of metabolic pathways and illuminates expected techniques for treating metabolic infections.

1.5 Moral Contemplations: Outlining the Course with Liability

As the translational excursion unfurls, the moral compass directing angelfish research turns out to be progressively vital. The dependable and moral treatment of angelfish, the straightforward correspondence of discoveries, and the thought of more extensive natural and protection morals are fundamental parts of the translational interaction. This moral supporting guarantees that the steps made in angelfish research are experimentally thorough as well as line up with rules that maintain the prosperity of both the species and the more extensive environment.

1.5.1 Creature Government assistance and Moral Treatment:

The moral treatment of angelfish in research is central. Moral contemplations include guaranteeing the prosperity of the creatures, limiting pressure, and carrying out empathetic practices in hereditary control and exploratory methods. As angelfish act as diplomats for natural investigation, their moral treatment isn't just a logical basic however an ethical obligation.

1.5.2 Informed Assent in Human Examinations:

For translational exploration to overcome any barrier from angelfish discoveries to human clinical investigations, it is vital to get educated assent. Moral contemplations include straightforwardly imparting the idea of the exploration, likely dangers, and advantages to human members. Informed assent guarantees that people know about the review's goals and enthusiastically take part in research that might impact their wellbeing.

1.5.3 Ecological and Preservation Morals:

Translational examination including angelfish ought to consider natural and preservation morals. Mindful practices incorporate limiting the effect on wild populaces, supporting protection endeavors, and executing feasible examination rehearses. Analysts ought to effectively participate in protection drives and add to the conservation of biological systems where angelfish species dwell.

1. **A brief overview of the significance of angelfish research in the biomedical field.**

   Angelfish research has arisen as a vital and dynamic field inside the biomedical space, contributing special experiences that rise above the limits of conventional model life forms. As mainstream researchers dives into the complexities of hereditary qualities, physiology, and infection components, angelfish (Pterophyllum spp.) has become the dominant focal point, offering a multilayered stage for investigation. This outline plans to enlighten the meaning of angelfish research in the biomedical field, disentangling its authentic setting, special attributes, and its significant ramifications for figuring out human wellbeing.

1.1 The Verifiable Woven artwork of Angelfish Exploration

The excursion of angelfish research follows its beginnings to the convergence of interest and logical request. At first enamoring aquarium aficionados with their enrapturing excellence, angelfish progressed from fancy charm to subjects of deliberate review. Early examinations essentially revolved around conduct perceptions, environment, and fundamental science.

Notwithstanding, it was the slow acknowledgment of angelfish's true capacity as a model organic entity that noticeable a groundbreaking movement.

During the twentieth 100 years, as sub-atomic science and hereditary exploration picked up speed, researchers started to see the value in the hereditary pliability of angelfish. This acknowledgment laid the preparation for precise lab tests, preparing for angelfish to rise above its job as an enlivening oceanic animal varieties to turn into a hereditarily manipulable creature with sweeping ramifications for biomedical examination.

1.2 The Hereditary Flexibility of Angelfish

1.2.1 A Living Outline:

At the core of angelfish research lies its hereditary outline. The sequencing of the angelfish genome has uncovered a thorough library of hereditary data, offering specialists uncommon experiences into the sub-atomic engineering of this species. The accessibility of a total genome gives an establishment to understanding quality capability, administrative components, and the hereditary premise of physiological cycles.

1.2.2 CRISPR/Cas9 Innovation: Accuracy in Hereditary Control:

The approach of CRISPR/Cas9 innovation has raised angelfish exploration higher than ever of accuracy. This progressive hereditary control instrument empowers specialists to adjust explicit qualities with unrivaled exactness. By presenting designated hereditary changes, researchers can unwind the useful meaning of individual qualities, mimic illness conditions, and investigate the hereditary underpinnings of intricate attributes.

1.3 Preservation of Natural Cycles: An Extension to Human Wellbeing

The meaning of angelfish research reaches out past its class and hereditary manipulability; it lies in the preservation of natural cycles that line up with those of people. Regardless of the transformative distance among angelfish and people, the essential pathways overseeing improvement, digestion, and illness reactions display astounding likenesses. This preservation of natural cycles gives a scaffold between angelfish research and the complexities of human wellbeing.

1.3.1 Metabolic Bits of knowledge:

Investigating the metabolic pathways in angelfish offers a window into the cell processes overseeing energy guideline. Bits of knowledge acquired from angelfish digestion studies contribute not exclusively to figuring out the physiological transformations of this species yet additionally hold suggestions for human metabolic problems.

As angelfish answer changes in their oceanic climate, the metabolic transformations noticed offer a remarkable point of view on how life forms explore ecological difficulties.

1.3.2 Cardiovascular Equals:

The cardiovascular framework, a life saver in both angelfish and people, has turned into a point of convergence of translational exploration. Studies investigating the complexities of angelfish cardiovascular turn of events and capability give bits of knowledge suggestions that reach out past oceanic domains. The equals among angelfish and human hearts offer a remarkable vantage point for figuring out cardiovascular infections and concocting novel helpful procedures.

1.4 Sickness Displaying and Helpful Experiences

Angelfish's job as a model organic entity reaches out past unwinding essential natural cycles; it fills in as a unique stage for sickness demonstrating and restorative investigation. By presenting hereditary changes that emulate human circumstances, scientists can make angelfish models of explicit infections, preparing for a more profound comprehension of illness instruments and the improvement of designated mediations.

1.4.1 Bits of knowledge into Neurological Problems:

The neurobiology of angelfish presents a charming road for figuring out neurological problems. Concentrating on the complexities of angelfish neuro-development, conduct, and mental capabilities offers bits of knowledge into equal cycles in the human mind. The likely applications in propelling comprehension we might interpret neurological issues and the advancement of designated mediations highlight the meaning of angelfish research in the neurological space.

1.4.2 Medication Revelation and Therapeutics:

Angelfish research adds to the medication revelation pipeline by filling in as a proving ground for drug specialists. The distinguishing proof of bioactive mixtures, investigation of novel medication targets, and the evaluation of medication viability in angelfish models offer a unique stage for making an interpretation of disclosures into unmistakable clinical arrangements. Angelfish, in this unique circumstance, turns into a basic partner in the mission for growing new helpful mediations.

1.5 Immunology and Incendiary Reactions: Protecting Equilibrium

As the resistant framework coordinates the sensitive harmony among safeguard and resilience, angelfish research gives an exceptional focal point through which to concentrate on immunology and incendiary reactions.

Bits of knowledge acquired from angelfish studies contribute not exclusively to figuring out essential safe systems yet additionally to unwinding the intricacies of immune system problems. The possible headways in immunotherapy enlivened by angelfish research hold guarantee for changing the scene of human safe related illnesses.

1.6 Moral Contemplations and Ecological Awareness

The moral components of angelfish research are necessary to its importance in the biomedical field. Mindful treatment of angelfish, straightforward correspondence of discoveries, and contemplations of more extensive ecological and preservation morals guarantee that the steps made in angelfish research line up with rules that maintain the prosperity of both the species and the more extensive biological system.

1.6.1 Creature Government assistance:

Moral contemplations include guaranteeing the prosperity of angelfish subjects in research settings. Limiting pressure, executing sympathetic practices in hereditary control, and keeping up with elevated requirements of creature government assistance are fundamental parts of capable logical investigation. The moral treatment of angelfish isn't just a logical basic however an ethical obligation that underlines the meaning of moral contemplations in angelfish research.

1.6.2 Protection Morals:

As angelfish are frequently gathered from their regular living spaces for research, protection morals assume a urgent part. Specialists are progressively aware of the ecological effect of their investigations, pushing for supportable exploration rehearses, and adding to preservation drives. Protection morals guarantee that the magnificence and biodiversity of angelfish territories are safeguarded for people in the future.

2. **Establishing the purpose of the book: translating angelfish findings to benefit human health.**

In the immense field of logical writing, the motivation behind a book fills in as its directing star, molding the story and outlining the talk. This investigation digs into the motivation behind a book devoted to the interpretation of angelfish discoveries, explaining the significant meaning of unwinding the insider facts held inside the oceanic domains of Pterophyllum spp. furthermore, the groundbreaking potential these disclosures bear for human wellbeing. The intention isn't simply to relate logical undertakings yet to wind around a convincing story that crosses the waters of revelation, development, and the interpretation of information from the aquarium to the center.

2. Setting the Stage: The One of a kind Material of Angelfish Exploration

Prior to revealing the reason, it is crucial for set the stage by valuing the uniqueness of angelfish research. From the rich influence of their balances to the complexities of their hereditary cosmetics, angelfish enthrall aquarium lovers as well as scientists looking to translate the natural orchestra that unfurls inside their oceanic world. The reason for this book is inherently attached to the acknowledgment that angelfish, frequently viewed as fancy oceanic occupants, harbor privileged insights that might actually alter how we might interpret human wellbeing.

3. The Reason Divulged

3.1 Connecting the Species Gap: A Translational Odyssey

The main role of this book is to set out on a translational odyssey that spans the apparent hole between the oceanic universe of angelfish and the multifaceted intricacies of the human body. While the developmental distance between these two species might appear to be immense, the hereditary, physiological, and conduct matches uncovered through angelfish research offer a remarkable extension for logical investigation. The book looks to explore this extension, unwinding the hereditary woven artwork of angelfish and making an interpretation of its discoveries into substantial bits of knowledge with direct ramifications for human wellbeing.

3.2 Illness Displaying and Restorative Experiences: Making ready for Accuracy Medication

At the center of the book's motivation lies the undertaking to outfit angelfish as a unique model for infection investigation. By digging into the hereditary and physiological subtleties of angelfish, scientists can make illness models that impersonate human circumstances. This translational jump takes into account the examination of sickness instruments, the testing of possible remedial intercessions, and the advancement of accuracy medication draws near. The reason, accordingly, isn't restricted to the aquarium however reaches out to the facility, where angelfish discoveries hold the possibility to reform sickness understanding and therapy techniques.

3.3 Cardiovascular Disclosures: Unwinding the Main issue

A critical feature of the book's motivation is to enlighten the cardiovascular equals among angelfish and people. By exploring the complexities of angelfish cardiovascular turn of events and capability, analysts can acquire experiences that stretch out past the sea-going domain. The object is to unwind the main issue at hand - figuring out cardiovascular illnesses and contriving novel remedial systems. Angelfish, with its monitored cardiovascular cycles, turns into a living material on which the brushstrokes of disclosure might illustrate human heart wellbeing.

3.4 Neurological Disclosures: Unraveling the Angelfish Cerebrum and Then some

The book's motivation reaches out to the domains of neurobiology, investigating the perplexing dance of neurons inside the angelfish mind. Through the investigation of neurodevelopment, conduct, and mental capabilities in angelfish, analysts can open neurological disclosures with suggestions for human neuroscience. The object is to disentangle the angelfish cerebrum and, likewise, add to how we might interpret neurological problems. From aquarium tanks to the neuroscience lab, the excursion is one of interpretation and change.

3.5 Medication Disclosure and Therapeutics: From Tank to Center

Chasing after translational importance, the book expects to exhibit angelfish as something other than amphibian colleagues. It positions angelfish as partners in the medication revelation process. By recognizing bioactive mixtures, investigating novel medication targets, and testing drug specialists inside the watery limits of the angelfish tank, the object is to show the capability of these discoveries to progress from the tank to the center. Angelfish, in this specific situation, arise as accomplices in the journey for novel helpful mediations.

3.6 Immunology and Provocative Reactions: Watching the Equilibrium

The reason reaches out into the domain of immunology, where angelfish with their complicated invulnerable frameworks become subjects of investigation. By unwinding the intricacies of angelfish immunology and provocative reactions, scientists mean to contribute not exclusively to the comprehension of essential resistant instruments yet additionally to bits of knowledge into immune system issues. The design is to monitor the sensitive harmony between the safe framework's protection and resistance components, with suggestions for propelling immuno-therapy methodologies in human wellbeing.

3.7 Metabolic Pathways and Illnesses: Exploring the Phone Landscape

Exploring the phone territory of metabolic pathways is a significant reason for the book. Angelfish, with its transformations to energy guideline and reactions to natural changes, turns into a living guide for grasping metabolic cycles. The intention is to investigate the ramifications of angelfish research for human metabolic problems, giving experiences that might direct techniques for treating conditions connected with energy digestion.

4. Translational Difficulties and Moral Contemplations

The reason for the book isn't heedless to the difficulties that go with the translational excursion. By tending to translational difficulties head-on, including species-explicit contrasts, natural awareness, and moral contemplations, the book plans to direct specialists through the intricacies.

The intention isn't just to feature victories however to furnish established researchers with the apparatuses and moral compass important to capably explore the translational way.

Chapter 1

The Angelfish Model

The Angelfish Model addresses a change in outlook in logical investigation, as specialists dive into the multifaceted universe of angelfish to reveal mysteries that might upset human wellbeing. This dazzling species, eminent for its dynamic tones and smooth developments, is something other than a sea-going exhibition — it holds the way to opening a bunch of organic secrets. This far reaching investigation will dig into the different aspects of the Angelfish Model, from its regenerative capacities and resistant framework complexities to the genomic scene that might prepare for imaginative forward leaps in clinical science.

Angelfish Physiology and Recovery:

One of the most spellbinding parts of the Angelfish Model is its surprising regenerative capacities. Not at all like people, angelfish have the ability to recover harmed tissues with bewildering effectiveness. Researchers are fastidiously concentrating on the cell and atomic cycles that underlie this regenerative ability, expecting to unravel the hereditary code liable for such amazing accomplishments. By understanding these systems, scientists imagine the likely utilization of angelfish-propelled regenerative techniques in human medication, offering new expectation for people with wounds or degenerative illnesses.

Insusceptible Framework Bits of knowledge:

The insusceptible arrangement of angelfish remains as a fortification against a different exhibit of microbes. Researching the subtleties of angelfish invulnerable reactions gives important bits of knowledge into the development of resistant guards. Scientists are acutely looking at the exceptional transformations that permit angelfish to successfully avoid infections. The expectation is that disentangling the insider facts of angelfish insusceptibility will prompt the advancement of novel immunotherapies and immunizations for people, supporting our guards against irresistible illnesses and tumors.

Genomic Investigation:

At the core of the Angelfish Model lies a mother lode of hereditary data. Researchers are directing exhaustive genomic studies to unwind the particular qualities liable for the regenerative ability, resistant flexibility, and by and large versatility of angelfish. The genomic scene of angelfish fills in as a plan for likely helpful focuses in human wellbeing.

This inside and out investigation might make ready for customized medication, taking into consideration custom-made medicines in view of a person's hereditary profile and tending to hereditary issues with freshly discovered accuracy.

Illustrations in Versatility:

Angelfish flourish in different natural circumstances, exhibiting a great degree of flexibility. Analysts are examining the hereditary and physiological variables that add to this versatility, expecting to comprehend how angelfish explore natural difficulties. The Angelfish Model gives a novel viewpoint on hereditary transformations that might move procedures to upgrade human flexibility and versatility notwithstanding ecological and wellbeing challenges.

Interpreting Angelfish Discoveries for Human Wellbeing:

A definitive objective of Angelfish research is to make an interpretation of its discoveries into unmistakable advantages for human wellbeing. As researchers disentangle the secrets of angelfish science, they are effectively making progress toward creating useful applications. These applications length a wide range, from regenerative medication and immunotherapy to the improvement of designated treatments for hereditary problems. The Angelfish Model fills in as an extension between sea life science and clinical science, offering an abundance of chances to propel human prosperity.

Regenerative Medication: Overcoming any issues:

Angelfish, with their excellent regenerative capacities, offer an outline for regenerative medication in people. By understanding the hereditary and cell processes that drive tissue recovery in angelfish, analysts are investigating ways of applying these standards to human medication. From mending wounds to tending to degenerative infections, the experiences acquired from the Angelfish Model hold monstrous potential for progressing regenerative medication and working on understanding results.

Immunotherapy Advancements:

The investigation of angelfish safe frameworks is making ready for imaginative immunotherapies. Angelfish, with their vigorous invulnerable guards, give a characteristic wellspring of motivation for growing new ways to deal with support the human insusceptible framework. This incorporates the investigation of novel immunization methodologies and immunomodulatory treatments that tackle the examples gained from angelfish. The Angelfish Model might add to a change in outlook by they way we approach irresistible sicknesses and malignant growth therapy, prompting more powerful and designated mediations.

Antibodies: Gaining from Nature's Defenders:

Angelfish, as gatekeepers of their submerged domains, have advanced multifaceted guard instruments against microbes. Analysts are acutely noticing the invulnerable reactions of angelfish to distinguish key parts that add to their antibody like security. This information is important in the advancement of immunizations for people, with the possibility to improve our capacity to forestall and battle irresistible sicknesses. The Angelfish Model opens new roads for immunization research, giving experiences into the plan of more vigorous and versatile antibody procedures.

Genomic Focuses for Therapeutics:

Genomic investigation of angelfish is revealing a rich embroidery of hereditary data. Researchers are carefully distinguishing explicit qualities related with regenerative capacities, safe versatility, and flexibility. These qualities act as likely focuses for restorative mediations in human wellbeing. The Angelfish Model might add to the improvement of quality treatments, accuracy medication draws near, and creative medicines for hereditary issues. By translating the genomic language of angelfish, scientists are establishing the groundwork for another period of focused on and customized therapeutics.

Customized Medication: Fitting Medicines to People:

The Angelfish Model envoys another time in customized medication. As scientists uncover the hereditary varieties that add to angelfish versatility and flexibility, they imagine a future where clinical medicines are customized to individual hereditary profiles. This customized approach holds guarantee for additional successful and designated treatments, limiting antagonistic impacts and improving treatment results. The Angelfish Model isn't simply a concentrate in sea life science; it is a guide for propelling medication towards accuracy and individualized care.

Moral Contemplations in Angelfish Exploration:

While the Angelfish Model offers colossal potential for propelling human wellbeing, moral contemplations are vital. Specialists should explore the fragile harmony between logical disclosure and mindful examination rehearses. Guaranteeing the prosperity of angelfish populaces and biological systems is critical, as the quest for information ought to be joined by a pledge to ecological stewardship. Moral structures in Angelfish research include straightforward correspondence, protection endeavors, and capable utilization of logical discoveries to stay away from unseen side-effects.

Cooperative Endeavors: Crossing over Disciplines for Progress:

The Angelfish Model highlights the significance of interdisciplinary joint effort. Researchers from different fields, including sea life science, hereditary qualities, immunology, and medication, are meeting up to disentangle the intricacies of angelfish science. Cooperative endeavors improve the extension and profundity of examination, taking into consideration a more exhaustive comprehension of the Angelfish Model. The collaboration between disciplines speeds up the interpretation of discoveries into pragmatic applications, encouraging a comprehensive way to deal with propelling human wellbeing.

Instructive Effort and Public Commitment:

The Angelfish Model presents a chance for instructive effort and public commitment. Imparting the meaning of this examination to the more extensive public encourages understanding and backing. Instructive drives can rouse future researchers and progressives, underscoring the interconnectedness of marine environments and human wellbeing. Public commitment likewise energizes a feeling of shared liability regarding the moral and supportable direct of Angelfish research, guaranteeing that the advantages got from this model are offset with a pledge to natural protection.

Difficulties and Future Bearings:

While the Angelfish Model holds massive commitment, it isn't without its difficulties. Specialists face obstacles in translating the mind boggling hereditary code, grasping complex physiological cycles, and guaranteeing the moral treatment of angelfish populaces. Future headings in Angelfish research include defeating these difficulties through mechanical headways, moral systems, and global joint effort. The excursion ahead incorporates refining how we might interpret angelfish science, making an interpretation of discoveries into unmistakable applications, and addressing arising moral contemplations to expand the positive effect on human wellbeing.

Natural Ramifications:

The investigation of the Angelfish Model goes past the domains of human wellbeing; it has more extensive natural ramifications. Angelfish possess fragile marine biological systems, and examination exercises should think about the likely environmental effects. Manageable exploration rehearses, living space protection, and mindful trial and error are fundamental to shielding the normal natural surroundings of angelfish and other marine species. By coordinating natural contemplations into Angelfish research, researchers add to the safeguarding of biodiversity and the drawn out soundness of marine biological systems.

Worldwide Effect and Global Cooperation:

The Angelfish Model can possibly have a worldwide effect on human wellbeing, rising above topographical limits. Global joint effort is urgent for pooling assets, ability, and different points of view. By cultivating associations between research establishments, states, and associations around the world, researchers can speed up the interpretation of Angelfish discoveries into useful applications. The worldwide effect of the Angelfish Model highlights the interconnected idea of logical advancement and the significance of aggregate endeavors in propelling human wellbeing.

Public Arrangement and Guideline:

As Angelfish research advances, the improvement of capable public arrangements and guidelines becomes basic. Legislatures and administrative bodies assume a critical part in guaranteeing that Angelfish research sticks to moral norms, natural manageability, and public wellbeing. The foundation of rules for the moral treatment of angelfish in research settings, preservation measures, and the mindful utilization of hereditary data is fundamental for outfitting the maximum capacity of the Angelfish Model while relieving likely dangers.

Cultural Insight and Acknowledgment:

The outcome of Angelfish research is complicatedly attached to cultural insight and acknowledgment. Straightforward correspondence about the objectives, strategies, and expected advantages of this examination is crucial for earn public help. Tending to worries and cultivating a positive view of Angelfish research guarantees that society embraces the expected progressions in human wellbeing while at the same time perceiving the significance of moral contemplations and ecological protection.

Financial Ramifications:

The Angelfish Model, with its true capacity for momentous revelations, may have huge monetary ramifications. The improvement of new clinical medicines, treatments, and advances enlivened by angelfish science can add to monetary development and occupation creation. Furthermore, the biotechnology and drug businesses might encounter a flood in development because of experiences acquired from the Angelfish Model. Understanding the monetary elements of Angelfish research is crucial for policymakers, financial backers, and ventures to capably bridle its maximum capacity.

1.1 Overview of angelfish species used in biomedical research.

Angelfish, a group of dynamically hued and effortlessly swimming fish, have risen above their status as aquarium top choices to become vital supporters of biomedical examination.

This complete investigation digs into the assorted types of angelfish used in biomedical examinations, stressing their remarkable qualities and the important bits of knowledge they accommodate propelling human wellbeing. From regenerative capacities and resistant framework elements to hereditary flexibility, every angelfish species divulges an unmistakable arrangement of natural credits that specialists influence to open the insider facts of clinical science.

Pterophyllum Scalare: The Famous Freshwater Angelfish

Pterophyllum scalare, regularly known as the freshwater angelfish, is a notorious species worshipped for its striking appearance and particular three-sided shape. Biomedical scientists tackle the regenerative capability of P. scalare to comprehend tissue fix components. Concentrates on the regenerative capacities of this species offer experiences into cell expansion, separation, and tissue renovating. By unraveling the hereditary and sub-atomic underpinnings of P. scalare's regenerative ability, specialists expect to foster systems for upgrading tissue recovery in people, possibly changing the field of regenerative medication.

Genomic Bits of knowledge from Pterophyllum Altum: Unwinding the Hereditary Code

Pterophyllum altum, ordinarily alluded to as the altum angelfish, stands apart for its magnificent height and perplexing balance designs. Biomedical analysts center around the genomic scene of P. altum to translate the hereditary premise of angelfish versatility and flexibility. The perplexing investigation of its genome gives a diagram to distinguishing explicit qualities related with insusceptible reactions, natural versatility, and hereditary variety. The genomic bits of knowledge got from P. altum make ready for designated therapeutics, customized medication, and a more profound comprehension of human hereditary qualities.

Regenerative Marvels of Pterophyllum Leopoldi: The Bantam Angelfish

Pterophyllum leopoldi, or the bantam angelfish, is prestigious for its conservative size and lively shading. This species becomes the dominant focal point in regenerative medication research because of its capacity to fix harmed tissues effectively. Researchers explore the cell and atomic components that empower P. leopoldi to recover tissues with accuracy. Understanding the regenerative marvels of the bantam angelfish might hold the way to creating imaginative treatments for human patients experiencing wounds, degenerative infections, or organ harm.

Safe Framework Versatility in Holacanthus Ciliaris: The Sovereign Angelfish

Holacanthus ciliaris, normally known as the sovereign angelfish, flaunts a great appearance with dynamic blue and yellow shades. Biomedical specialists center around the insusceptible framework flexibility of H. ciliaris to unwind the complexities of insusceptible reactions. This species displays powerful guards against various microorganisms, making it a significant model for concentrating on immunological cycles. Bits of knowledge acquired from the sovereign angelfish's safe framework might prompt the improvement of cutting edge immunotherapies, immunizations, and a more profound comprehension of the human resistant reaction.

Chaetodontoplus Septentrionalis: The Ocean Goldie and Hereditary Revelations

Chaetodontoplus septentrionalis, generally known as the ocean goldie, spellbinds with its brilliant shades and particular markings. Specialists go to this species for its hereditary versatility and flexibility. By leading genomic concentrates on C. septentrionalis, researchers intend to distinguish qualities related with hereditary variety and ecological variation. These hereditary revelations have sweeping ramifications, offering likely focuses for treating hereditary issues, upgrading versatility, and propelling the field of customized medication.

Pomacanthus Imperator: Sovereign Angelfish and Bits of knowledge into Natural Transformation

Pomacanthus imperator, otherwise called the sovereign angelfish, rules with its striking shading and unmistakable examples. Analysts investigate the hereditary and physiological elements that add to the head angelfish's flexibility to different ecological circumstances. By understanding the systems that empower this species to flourish in various living spaces, researchers gain significant bits of knowledge into ecological transformation. The examples gained from P. imperator may illuminate procedures for improving human versatility and strength notwithstanding changing natural circumstances.

Moral Contemplations in Angelfish Biomedical Exploration

As angelfish species become essential to biomedical exploration, moral contemplations become the dominant focal point. Specialists should explore the fragile harmony between logical revelation and mindful treatment of these amphibian species. Moral systems in angelfish research include straightforward correspondence, preservation endeavors, and capable utilization of logical discoveries to keep away from potentially negative side-effects. The government assistance of angelfish populaces and biological systems should be focused on to guarantee the moral lead of examination and the supportability of these significant models.

Cooperative Endeavors: Crossing over Disciplines for Biomedical Advancement

The use of angelfish species in biomedical exploration highlights the significance of interdisciplinary coordinated effort. Researchers from different fields, including sea life science, hereditary qualities, immunology, and medication, meet to disentangle the intricacies of angelfish science. Cooperative endeavors upgrade the extension and profundity of examination, taking into consideration a more thorough comprehension of these oceanic models. The collaboration between disciplines speeds up the interpretation of discoveries into commonsense applications, encouraging an all encompassing way to deal with propelling human wellbeing.

Instructive Effort and Public Commitment to Angelfish Biomedical Exploration

The consideration of angelfish species in biomedical examination presents a chance for instructive effort and public commitment. Imparting the meaning of this exploration to the more extensive public encourages understanding and backing. Instructive drives can move future researchers and progressives, underscoring the interconnectedness of marine biological systems and human wellbeing. Public commitment likewise energizes a feeling of shared liability regarding the moral and feasible lead of angelfish biomedical exploration, guaranteeing that the advantages got from these models are offset with a pledge to ecological preservation.

Difficulties and Future Headings in Angelfish Biomedical Exploration

While angelfish species offer enormous potential for propelling human wellbeing, specialists face different difficulties. Translating the mind boggling hereditary code, grasping complex physiological cycles, and guaranteeing the moral treatment of angelfish populaces are among the obstacles. Future bearings in angelfish biomedical examination include beating these difficulties through mechanical progressions, moral structures, and global coordinated effort. The excursion ahead incorporates refining how we might interpret angelfish science, making an interpretation of discoveries into substantial applications, and addressing arising moral contemplations to expand the positive effect on human wellbeing.

Ecological Ramifications of Angelfish Biomedical Exploration

The usage of angelfish species in biomedical exploration goes past the lab, with more extensive ecological ramifications. Angelfish possess fragile marine biological systems, and examination exercises should think about the likely natural effects. Economical exploration rehearses, living space preservation, and mindful trial and error are fundamental to shielding the regular natural surroundings of angelfish and other marine species. By incorporating natural contemplations into angelfish biomedical examination, researchers add to the protection of biodiversity and the drawn out wellbeing of marine biological systems.

Worldwide Effect and Global Coordinated effort in Angelfish Biomedical Exploration

Angelfish biomedical exploration can possibly have a worldwide effect on human wellbeing, rising above geological limits. Worldwide coordinated effort is urgent for pooling assets, mastery, and different viewpoints. By encouraging associations between research foundations, states, and associations around the world, researchers can speed up the interpretation of angelfish discoveries into functional applications. The worldwide effect of angelfish biomedical examination highlights the interconnected idea of logical advancement and the significance of aggregate endeavors in propelling human wellbeing.

Public Strategy and Guideline in Angelfish Biomedical Exploration

As angelfish biomedical examination advances, the improvement of capable public approaches and guidelines becomes basic. States and administrative bodies assume a significant part in guaranteeing that angelfish research sticks to moral guidelines, ecological maintainability, and public security. The foundation of rules for the moral treatment of angelfish in research settings, protection measures, and the dependable utilization of hereditary data is fundamental for saddling the maximum capacity of these models while alleviating possible dangers.

Cultural Insight and Acknowledgment of Angelfish Biomedical Exploration

The outcome of angelfish biomedical examination is complicatedly attached to cultural discernment and acknowledgment. Straightforward correspondence about the objectives, techniques, and likely advantages of this exploration is crucial for earn public help. Tending to worries and cultivating a positive impression of angelfish biomedical exploration guarantees that society embraces the expected progressions in human wellbeing while at the same time perceiving the significance of moral contemplations and ecological protection.

Monetary Ramifications of Angelfish Biomedical Exploration

The usage of angelfish species in biomedical exploration might have huge monetary ramifications. The improvement of new clinical medicines, treatments, and innovations propelled by angelfish science can add to monetary development and occupation creation. Moreover, the biotechnology and drug ventures might encounter a flood in development because of experiences acquired from angelfish biomedical exploration. Understanding the financial elements of this exploration is imperative for policymakers, financial backers, and ventures to capably outfit its maximum capacity.

1.2 Unique characteristics that make angelfish a valuable model organism.

Angelfish, with their entrancing excellence and elegant developments, play rose above their part as decorative aquarium fish to become important model creatures in logical exploration. These remarkable oceanic animals have a heap of qualities that make them especially appropriate for examinations crossing different logical disciplines. From regenerative capacities and safe framework elements to genomic versatility, angelfish offer an abundance of experiences that add to headways in biomedical, natural, and hereditary examination.

Regenerative Ability:

Angelfish show surprising regenerative capacities, making them an ideal model for concentrating on tissue fix and recovery. Not at all like people, angelfish can proficiently recover harmed tissues, including balances and scales. Specialists concentrate on the cell and sub-atomic cycles engaged with this regenerative ability to comprehend the systems that might actually be tackled for regenerative medication. Bits of knowledge acquired from angelfish regenerative capacities can possibly reform medicines for wounds, organ harm, and degenerative sicknesses in people.

Various Species for Near Examinations:

The angelfish family includes a different cluster of animal categories, each with its own exceptional qualities. Scientists can use this variety for relative examinations, permitting them to investigate varieties in regenerative abilities, safe reactions, and genomic scenes among various species. This near approach gives an extensive comprehension of the basic natural systems and works with the distinguishing proof of shared traits and one of a kind transformations inside the angelfish family.

Genomic Strength:

Angelfish species show hereditary strength, permitting them to flourish in assorted natural circumstances. This flexibility is specifically noteworthy to geneticists and specialists concentrating on variation. By disentangling the hereditary parts that add to angelfish versatility, researchers gain experiences into the systems that empower living beings to adapt to evolving conditions. These hereditary bits of knowledge might have suggestions for grasping human hereditary variety, variation, and helplessness to natural elements.

Immunological Bits of knowledge:

The invulnerable arrangement of angelfish is a subject of serious exploration because of its strong safeguards against different microbes. Concentrating on the insusceptible reactions of angelfish gives important bits of knowledge into the advancement of safe instruments. Specialists can explore the elements that add to angelfish insusceptibility, possibly prompting the improvement of novel immunotherapies, immunizations, and a more profound comprehension of the human safe framework's intricacies.

Conduct Perceptions:

Angelfish show complex social ways of behaving and associations, making them a captivating model for conduct studies. Specialists can notice and break down friendly progressive systems, mating ways of behaving, and regional elements among angelfish. These conduct studies contribute not exclusively to how we might interpret fish conduct yet in addition offer experiences into more extensive parts of creature discernment and social designs.

Preservation Significance:

As agents of sensitive marine environments, angelfish species hold critical preservation significance. Concentrating on angelfish in their regular natural surroundings gives bits of knowledge into the effect of ecological changes on these creatures. Preservation endeavors can profit from understanding the variables affecting angelfish populaces and environments, adding to more extensive marine protection drives.

Amphibian Models for Ecological Variation:

Angelfish flourish in different sea-going conditions, from freshwater to coral reefs. Specialists concentrating on ecological variation can use angelfish as models to examine how these organic entities adjust to various water conditions. The versatile techniques utilized by angelfish might offer bits of knowledge into the flexibility of oceanic species even with ecological changes, illuminating preservation practices and biological system the executives.

Reproducibility and Short Age Times:

Angelfish display somewhat short age times and high regenerative rates, working with trial studies and perceptions over various ages. This trademark is favorable for scientists concentrating on hereditary legacy, formative cycles, and the impacts of natural variables across progressive ages. The capacity to rapidly create information adds to the reproducibility and strength of logical discoveries.

Openness for Lab Studies:

Angelfish are appropriate for research facility studies, as they adjust to controlled conditions and replicate promptly in bondage. This openness makes them advantageous models for scientists leading investigations on different parts of their science, including hereditary qualities, conduct, and physiology. The controlled research facility setting improves the accuracy and unwavering quality of trial information.

Instructive Worth:

Angelfish, with their enthralling appearance and different ways of behaving, hold instructive worth in both scholar and public settings. They act as connecting with models for understudies and the overall population, encouraging interest in sea life science, hereditary qualities, and natural science. Instructive effort programs frequently use angelfish to convey logical ideas and advance mindfulness about the significance of marine environments.

1.3 Historical context and evolution of angelfish research.

Angelfish, with their radiant varieties and unmistakable elegance, have charmed aquarium devotees as well as logical analysts trying to disentangle the secrets of these amphibian miracles. The verifiable setting and development of angelfish research mirror an excursion from simple profound respect of their style to a profound comprehension of their organic intricacies. This investigation traverses different logical disciplines, enveloping sea life science, hereditary qualities, immunology, and regenerative medication.

Early Experiences and Aquarium Interest:

The interest with angelfish goes back hundreds of years, with verifiable records reporting their presence in native societies and early human advancements. Local to the freshwater waterways and floods of South America, angelfish, especially the notorious Pterophyllum scalare, were experienced by voyagers and naturalists during the period of disclosure. Early portrayals in regular history delineations and tales feature the stylish charm of angelfish, establishing the groundwork for their ubiquity in aquariums.

In the late nineteenth and mid twentieth hundreds of years, as aquarium keeping acquired prevalence, angelfish became valued augmentations to oceanic showcases. Their unmistakably lengthened bodies and striking blade setups made them sought-after examples in the arising universe of fancy fishkeeping. This period denoted the start of the crossing point between the stylish enthusiasm for angelfish and their job as subjects of logical interest.

Development of Angelfish as Model Living beings:

As the twentieth century advanced, established researchers perceived the capability of angelfish as model living beings for different exploration tries. The foundation of angelfish rearing projects in bondage worked with controlled tests and perceptions. The unmistakable benefit of angelfish in research center settings, with their flexibility to bondage and moderately short age times, made them helpful for a great many examinations.

Hereditary Investigations and Reproducing Projects:

One critical achievement in the development of angelfish research was the rise of hereditary examinations and reproducing programs. The mid-twentieth century saw a flood in interest among geneticists investigating legacy designs, variety varieties, and morphological qualities in angelfish. Reproducing programs pointed toward creating explicit variety varieties and blade arrangements became research centers for concentrating on hereditary legacy, laying the preparation for grasping the hereditary premise of these qualities.

Through particular rearing, scientists clarified the standards of Mendelian hereditary qualities, unwinding the innate instruments administering the transmission of characteristics starting with one age then onto the next. The information acquired from these hereditary examinations not just added to the field of fish hereditary qualities yet in addition had suggestions for understanding more extensive hereditary standards relevant to different organic entities, including people.

Immunological Examinations:

Angelfish, with their complicated safe frameworks, became subjects of immunological examinations in the last 50% of the twentieth 100 years. The strong safe reactions saw in angelfish, especially in species like Holacanthus ciliaris, ignited interest among immunologists looking to grasp the components behind their microorganism opposition.

Concentrates on angelfish resistant reactions gave bits of knowledge into the development of safe frameworks in amphibian creatures. The distinguishing proof of explicit safe pathways, cytokines, and the job of mucosal resistance in angelfish offered equals to earthly vertebrates. Immunologists perceived the capability of angelfish as models for concentrating on have microbe communications and creating novel immunotherapies.

Regenerative Medication and Tissue Fix:

In the domain of regenerative medication, angelfish arose as trailblazers in the investigation of tissue fix and recovery. The novel capacity of angelfish, for example, Pterophyllum leopoldi, to recover harmed balances and tissues turned into a point of convergence for specialists looking to open the insider facts of regenerative cycles.

Trial concentrates on the cell and sub-atomic components of tissue recovery in angelfish gave essential information to regenerative medication research. The recognizable proof of key flagging pathways, foundational microorganism elements, and regenerative signs in angelfish has suggestions for creating remedial procedures to improve tissue fix in people. This road of exploration can possibly change the field of regenerative medication, offering new experiences into wound recuperating, organ recovery, and tissue designing.

Genomic Investigation and Transformative Experiences:

Headways in genomic advancements in the late twentieth and mid 21st hundreds of years moved angelfish examination into the domain of genomics. The unwinding of the angelfish genome, including species like Pterophyllum altum and Chaetodontoplus septentrionalis, shed light on the hereditary engineering hidden their extraordinary attributes and transformations.

Genomic investigation permitted scientists to follow the transformative history of angelfish, giving bits of knowledge into their phylogenetic connections and uniqueness. Similar genomics across various angelfish species offered a nuanced comprehension of hereditary variety, transformation to different conditions, and potential specific tensions molding their genomes.

Interdisciplinary Coordinated effort and All encompassing Methodologies:

The development of angelfish research is set apart by a remarkable shift toward interdisciplinary coordinated effort. Scientists from fields as different as sea life science, hereditary qualities, immunology, and regenerative medication started teaming up to investigate the all encompassing parts of angelfish science. This synergistic methodology improved the profundity and expansiveness of angelfish research, cultivating a far reaching comprehension of their science according to various points of view.

Cooperative endeavors likewise reached out to worldwide organizations, with specialists from different nations pooling assets and mastery to boost the effect of angelfish research on a worldwide scale. The sharing of information, procedures, and assets sped up the speed of revelations and expanded the uses of angelfish research across topographical limits.

Moral Contemplations and Protection Mindfulness:

As angelfish research extended, moral contemplations in regards to the treatment of these living beings acquired conspicuousness. Moral systems were laid out to guarantee the compassionate and capable lead of analyses, stressing the prosperity of angelfish populaces.

Protection mindfulness turned into a fundamental part of angelfish research, perceiving the significance of saving their regular natural surroundings and adding to more extensive marine preservation endeavors.

Specialists effectively participated in instructive effort to bring issues to light about the moral elements of angelfish research and the meaning of marine protection. Public commitment drives focused on aquarium lovers, understudies, and the overall population encouraged a comprehension of the interconnectedness between angelfish research, moral contemplations, and natural stewardship.

Future Headings and Worldwide Effect:

As angelfish research keeps on developing, future headings highlight a considerably more coordinated and effective logical scene. Progressions in advancements, for example, CRISPR-Cas9 quality altering hold the possibility to additionally unwind the hereditary complexities of angelfish, considering designated changes and exact examinations.

The worldwide effect of angelfish research reaches out past the domains of logical disclosure. Experiences acquired from concentrates on regenerative medication, hereditary qualities, immunology, and natural transformation have suggestions for human wellbeing, biotechnology, and biological preservation. Angelfish, once

respected for their stylish allure, have become representatives of logical advancement with sweeping ramifications for assorted fields.

Chapter 2

Genetics and Genomic Insights

Hereditary qualities and genomics, the many-sided dialects recorded in the DNA, have been the key part of figuring out life's plan. In the excursion of logical investigation, the clarification of hereditary qualities and the disentangling of genomic complexities stand as great accomplishments. This far reaching investigation digs into the universe of hereditary qualities and genomics, from the primary standards of heredity to the state of the art innovations forming how we might interpret the genomic scene. As we explore the domains of qualities, chromosomes, and the transformative embroidery encoded in DNA, we set out on an excursion that holds guarantee for reforming medication, horticulture, and our comprehension of life itself.

1. **Groundworks of Hereditary qualities: The Mendelian Inheritance**

   The account of hereditary qualities starts with the basic work of Gregor Mendel, the dad of current hereditary qualities. During the nineteenth hundred years, Mendel led earth shattering tests with pea plants, revealing the standards of heredity. His fastidious perceptions of attributes, for example, seed tone and bloom position laid the basis for how we might interpret predominant and passive characteristics, hereditary legacy, and the idea of alleles.

   1.1 Mendel's Regulations: Foundations of Heredity

   Mendel's regulations — the Law of Isolation and the Law of Free Variety — turned into the foundations of old style hereditary qualities. The Law of Isolation directs that every individual has two alleles for a given characteristic, and these alleles isolate during gamete development, guaranteeing every gamete conveys just a single allele. The Law of Free Arrangement expresses that alleles for various attributes isolate autonomously during the development of gametes, adding to the variety of hereditary blends in posterity.

1.2 From Pea Plants to Human Legacy

Mendel's standards reached out past pea plants, tracking down application in grasping human legacy. The coming of the chromosomal hypothesis of legacy in the mid twentieth 100 years, combined with progressions in microscopy, associated Mendel's regulations to the actual designs of chromosomes. The acknowledgment that qualities, the units of heredity, are situated on chromosomes denoted a urgent second in the combination of traditional hereditary qualities and cytology.

2. Chromosomes and the Atomic Dance of DNA

As innovation advanced, specialists dove further into the atomic domain, finding the particle that conveys the directions forever — deoxyribonucleic corrosive, or DNA. The clarification of the DNA structure by James Watson and Francis Cramp in 1953, with critical information given by Rosalind Franklin and Maurice Wilkins, denoted a turning point throughout the entire existence of hereditary qualities.

2.1 DNA Design: The Twofold Helix Outline

The DNA atom, looking like a rich twofold helix, comprises of two long strands of nucleotides contorted around one another. The nucleotides, including a phosphate bunch, a sugar particle, and one of four nitrogenous bases (adenine, thymine, cytosine, or guanine), encode the hereditary data. The reciprocal base matching — adenine with thymine and cytosine with guanine — structures the rungs of the helical stepping stool, guaranteeing the reliable transmission of hereditary data during cell replication.

2.2 Replication and the Focal Creed of Atomic Science

DNA replication, a principal cycle in cell division, guarantees the loyal transmission of hereditary data to girl cells. The Focal Creed of Sub-atomic Science, verbalized by Francis Kink, frames the progression of hereditary data: from DNA to RNA to proteins. Record, the combination of RNA from DNA, happens in the cell core, while interpretation, the change of RNA into proteins, happens in the cytoplasm.

2.3 Chromosomes and Karyotyping: Planning the Hereditary Scene

Chromosomes, the consolidated designs of DNA and proteins, house the qualities that decide a living being's attributes. The human genome contains 23 sets of chromosomes, each conveying a large number of qualities. Karyotyping, a method that orchestrates chromosomes by size and construction, offers a visual guide of a person's hereditary scene. Variations in chromosome number or design can prompt hereditary issues, stressing the significance of figuring out chromosomal respectability.

3. Genomic Transformation: Translating the Book of Life

The turn of the 21st century saw a seismic change in hereditary qualities with the Human Genome Task (HGP), a cooperative undertaking pointed toward sequencing the whole human genome. Finished in 2003, the HGP denoted a vital second in the genomic upset, opening the succession of three billion DNA base coordinates that comprise the human genome.

3.1 The Human Genome Undertaking: Plan of Intricacy

The Human Genome Undertaking not just given a thorough guide of the human genome yet additionally catalyzed innovative progressions that sped up genomic research across species. The distinguishing proof of qualities related with infections, the investigation of hereditary varieties among people, and the disclosure of the non-coding locales of the genome divulged the mind boggling intricacy of our hereditary cosmetics.

3.2 Past the Human Genome: Similar Genomics

The genomic unrest rises above human science, stretching out to near genomics. By contrasting the genomes of various species, specialists gain bits of knowledge into developmental connections, shared hereditary components, and the transformations that support biodiversity. Relative genomics traverses the tree of life, from microorganisms to plants to creatures, disentangling the hereditary strings that weave the embroidery of presence.

4. **Genomic Variety and Variety: The Mosaic of Life**

Genomic variety, an innate part of the developmental cycle, is the aftereffect of hereditary transformations, recombination, and the amassing of hereditary varieties over the long run. Understanding genomic variety among people and populaces is urgent for disentangling the hereditary premise of qualities, helplessness to illnesses, and the versatile instruments that drive advancement.

4.1 Single Nucleotide Polymorphisms (SNPs): A Genomic Kaleidoscope

Single Nucleotide Polymorphisms (SNPs), varieties in a solitary nucleotide at a particular genomic position, structure the genomic kaleidoscope that adds to variety. SNPs act as hereditary markers, supporting the distinguishing proof of qualities related with infections and offering experiences into populace hereditary qualities. Far reaching affiliation studies (GWAS) influence SNPs to relate hereditary varieties with phenotypic attributes.

4.2 Duplicate Number Varieties (CNVs): Genomic Modifications

Duplicate Number Varieties (CNVs) include modifications in the quantity of duplicates of enormous DNA fragments, adding to genomic variety. CNVs can impact quality articulation, and their relationship with illnesses highlights their importance in grasping hereditary inclinations. Propels in genomic advancements, for example, exhibit relative genomic hybridization (aCGH) and cutting edge sequencing (NGS), empower the recognition and portrayal of CNVs with remarkable goal.

5. **Practical Genomics: Translating the Language of Qualities**

Utilitarian genomics tries to disentangle the elements of qualities, their administrative components, and the unique collaborations inside the cell hardware.

Innovations, for example, transcriptomics, proteomics, and metabolomics give an all encompassing perspective on quality articulation, protein blend, and metabolic pathways, revealing insight into the organization of natural cycles.

5.1 Transcriptomics: Divulging the RNA Orchestra

Transcriptomics inspects the sum of RNA particles delivered by a cell, giving experiences into quality articulation designs. Strategies like RNA sequencing (RNA-seq) offer exceptional goal in profiling the transcriptome. The distinguishing proof of coding and non-coding RNAs adds to how we might interpret cell capabilities, formative cycles, and the administrative organizations that oversee quality articulation.

5.2 Proteomics: The Language of Proteins

Proteomics digs into the tremendous scene of proteins, the atomic machines that execute cell capabilities. Mass spectrometry and other high-throughput procedures empower the recognizable proof and evaluation of proteins, disentangling their parts in flagging pathways, enzymatic exercises, and cell structures. The powerful idea of the proteome mirrors the intricacy of natural frameworks.

6. Epigenetics: The Orchestra Past DNA Groupings

Epigenetics, the investigation of heritable changes in quality capability that don't include adjustments in the DNA grouping, adds a perplexing layer to how we might interpret hereditary qualities. Epigenetic adjustments, like DNA methylation and histone acetylation, impact quality articulation and add to cell separation, advancement, and reactions to natural prompts. The blossoming field of epigenomics investigates the epigenetic scenes across tissues, formative stages, and infection states.

6.1 DNA Methylation: Methyl Imprints on the Hereditary Material

DNA methylation includes the expansion of methyl gatherings to DNA atoms, frequently connected with quality quieting. The examples of DNA methylation assume essential parts in managing quality articulation and keeping up with cell character. Distortions in DNA methylation are ensnared in different illnesses, including disease, featuring the significance of understanding epigenetic elements.

6.2 Histone Changes: Chromatin Orchestra Directors

Histone proteins, around which DNA is twisted to shape chromatin, go through different adjustments that impact chromatin construction and quality availability.

Histone acetylation, methylation, and phosphorylation comprise the chromatin ensemble that organizes quality articulation. Epigenetic drugs focusing on histone changes offer possible helpful roads for sicknesses portrayed by abnormal quality guideline.

7. **Genomic Medication: From Seat to Bedside**

The abundance of genomic experiences has made ready for extraordinary methodologies in medication, all in all known as genomic medication. The joining of genomics into clinical practice holds guarantee for customized diagnostics, designated treatments, and the clarification of hereditary elements affecting infection vulnerability and treatment reactions.

7.1 Pharmacogenomics: Fitting Medicines to Hereditary Varieties

Pharmacogenomics researches what a person's hereditary cosmetics means for their reaction to drugs. Grasping hereditary varieties in drug digestion, adequacy, and unfavorable responses empowers the fitting of medication regimens to individual patients. This customized approach improves treatment results and limits the dangers of unfavorable impacts, denoting a change in outlook in the area of pharmacology.

7.2 Accuracy Oncology: Genomic Directing Guides in Disease Treatment

In the domain of malignant growth treatment, accuracy oncology saddles genomic data to direct helpful choices. Genomic profiling of growths distinguishes explicit transformations and sub-atomic adjustments, empowering the remedy of designated treatments intended to restrain the sub-atomic drivers of disease. The time of accuracy oncology proclaims another wilderness in disease treatment, where treatments are custom-made to the special genomic scene of every patient's cancer.

8. **Moral Contemplations and Cultural Ramifications**

As the domain of hereditary qualities and genomics progresses, moral contemplations and cultural ramifications come to the very front. Issues like hereditary protection, the possible abuse of hereditary data, and the impartial conveyance of genomic advances raise moral quandaries that require insightful and comprehensive conversations. Guaranteeing dependable practices in research, clinical applications, and the spread of genomic data becomes principal in exploring the moral scene.

8.1 Hereditary Protection and Informed Assent

Hereditary data, with its possible ramifications for people and their families, raises worries about protection and assent. The approach of direct-to-buyer hereditary testing adds a layer of intricacy, as people get sufficiently close to their genomic information.

Finding some kind of harmony between enabling people with information and protecting their security highlights the significance of strong moral systems and informed assent processes.

8.2 Value in Genomic Medication: Spanning the Gap

The availability and reasonableness of genomic advances present provokes in guaranteeing evenhanded admittance to genomic medication. Variations in medical services access, hereditary testing accessibility, and the translation of genomic information feature the requirement for comprehensive methodologies that address financial and social variables. Spanning the genomic partition requires a purposeful work to advance variety in genomic exploration and medical services execution.

9. Future Outskirts: CRISPR, Manufactured Science, and Then some

The fate of hereditary qualities and genomics unfurls on the bleeding edge of mechanical advancement. CRISPR-Cas9, a progressive quality altering instrument, permits exact changes of DNA successions, opening roads for remedying hereditary problems, making hereditarily adjusted creatures, and unwinding the useful outcomes of explicit hereditary components. Manufactured science, an interdisciplinary field at the crossing point of science and designing, imagines the production of fake natural frameworks with applications going from medication to ecological maintainability.

9.1 CRISPR-Cas9: Accuracy Altering of the Hereditary Code

CRISPR-Cas9, hailed as a forward leap in hereditary designing, empowers designated changes of DNA successions with uncommon precision. This progressive apparatus holds guarantee for rectifying hereditary transformations, designing harvests for improved strength, and exploring the utilitarian jobs of explicit qualities. The moral contemplations encompassing the utilization of CRISPR innovation highlight the requirement for mindful and straightforward practices.

9.2 Manufactured Science: Upgrading Life's Outline

Manufactured science imagines the upgrade of organic frameworks for down to earth applications. From designed microorganisms for modern cycles to the production of manufactured cells with specially crafted capabilities, engineered science rises above the limits of conventional hereditary designing. The moral ramifications of manufactured science bring up issues about the mindful plan and arrival of engineered creatures into regular environments.

2.1Exploration of angelfish genetics and its parallels to human genetics.

The multifaceted dance of qualities, woven into the texture of life, rises above species limits, interfacing the different embroidery of living organic entities. Angelfish, with their dazzling magnificence underneath the water's surface, have become more than fancy animals in aquariums; they are vital participants in the logical investigation of hereditary qualities. This broad investigation digs into the universe of angelfish hereditary qualities, unwinding the equals among angelfish and human hereditary qualities. From the transformative strings that tight spot us to the secrets encoded in DNA, the investigation of angelfish hereditary qualities offers a remarkable focal point through which we can unravel the mysteries of our own hereditary legacy.

1. **Genomic Scene of Angelfish: An Ensemble of Variety**

The underpinning of understanding angelfish hereditary qualities lies in disentangling their genomic scene. The hereditary cosmetics of angelfish species, going from Pterophyllum scalare to Chaetodontoplus septentrionalis, divulges an orchestra of variety formed by development, transformation, and biological specialties.

1.1 Near Genomics: Following Transformative Strings

Near genomics, the investigation of hereditary similitudes and contrasts across species, gives a guide to following the transformative strings that interface angelfish to different organic entities, including people. The investigation of shared qualities, administrative components, and genomic structures offers experiences into the rationed components that have endured the trial of transformative time.

1.2 Hereditary Variety Among Angelfish Species

The different exhibit of angelfish species presents a rich embroidery of hereditary variety. Every species, adjusted to its particular territory and biological specialty, conveys a one of a kind arrangement of hereditary varieties that add to its endurance and regenerative achievement. Disentangling the hereditary code of various angelfish species opens windows into the versatile systems that have permitted these sea-going wonders to flourish in assorted conditions.

2. **Qualities and Attributes: Unraveling the Language of Legacy**

At the core of hereditary qualities lies the language of legacy, where qualities coordinate the orchestra of characteristics that characterize a creature. Angelfish, with their variety of varieties, blade designs, and ways of behaving, offer a material whereupon the standards of hereditary legacy can be painted.

2.1 Mendelian Legacy in Angelfish Characteristics

The standards of Mendelian legacy, found through the spearheading work of Gregor Mendel, track down reverberation in the legacy of characteristics among angelfish. Prevailing and latent alleles decide the declaration of attributes, for example, variety examples and balance shapes, mirroring the very hereditary rules that administer qualities in people and different organic entities.

2.2 Polygenic Characteristics and Complex Legacy

Past the straightforwardness of Mendelian legacy, angelfish hereditary qualities unwinds the intricacy of polygenic attributes. Characteristics affected by different qualities, for example, the complicated variety designs and unpretentious varieties in blade morphology, grandstand the polygenic idea of hereditary legacy. Concentrating on polygenic characteristics in angelfish adds to how we might interpret the hereditary engineering basic complex aggregates.

3. Genomic Advances: Looking into the Angelfish Genome

Headways in genomic advancements have been instrumental in unwinding the secrets encoded in the angelfish genome. From customary hereditary planning to cutting edge sequencing procedures, these devices offer a brief look into the outline of angelfish hereditary qualities.

3.1 Hereditary Planning and Linkage Investigation

Hereditary planning and linkage examination have been essential in distinguishing the area of qualities on angelfish chromosomes. By following the legacy of explicit qualities through ages, specialists can make hereditary guides that feature the locales of the genome related with specific aggregates. This planning fills in as a compass for additional investigation of the fundamental hereditary systems.

3.2 Cutting edge Sequencing (NGS): Deciphering the Angelfish Genome

The appearance of Cutting edge Sequencing (NGS) advancements has altered the investigation of angelfish hereditary qualities. Entire genome sequencing permits specialists to proficiently interpret the whole hereditary data of angelfish species. The far reaching experiences acquired from NGS enable researchers to investigate the complexities of quality guideline, utilitarian components, and developmental transformations inside the angelfish genome.

4. Protection Hereditary qualities: Defending Biodiversity

Understanding the hereditary variety inside angelfish populaces isn't just a logical undertaking yet in addition a pivotal part of preservation science. Protection hereditary qualities investigates the hereditary elements affecting populace elements, transformation to natural changes, and the drawn out feasibility of angelfish species.

4.1 Hereditary Markers and Populace Construction

Hereditary markers, for example, microsatellites and single nucleotide polymorphisms (SNPs), act as signs in the investigation of angelfish populaces. Examining these markers gives experiences into populace structures, relocation designs, and the effect of human exercises on hereditary variety. Protection endeavors can profit from a nuanced comprehension of hereditary variety to figure out systems for saving angelfish biodiversity.

4.2 Strength Qualities and Ecological Variation

The investigation of angelfish hereditary qualities reaches out to distinguishing flexibility qualities related with ecological variation. Examining the hereditary premise of qualities that give protection from natural stressors, like changes in water temperature or living space corruption, adds to preservation methodologies. Preservation hereditary qualities, entwined with biological contemplations, turns into an amazing asset for protecting angelfish populaces in their regular living spaces.

5. Equals to Human Hereditary qualities: Examples from the Amphibian

Domain

The equals between angelfish hereditary qualities and human hereditary qualities offer a charming story of shared organic standards. From the complexities of hereditary legacy to the effect of hereditary minor departure from characteristics and infections, the investigation of angelfish hereditary qualities turns into a focal point through which we can gather bits of knowledge into our own genomic legacy.

5.1 Transformative Protection of Qualities

The protection of specific qualities across transformative distances highlights the antiquated hereditary discoursed that associate angelfish to people. Qualities engaged with major organic cycles, like undeveloped turn of events, safe reaction, and metabolic pathways, show developmental protection. Concentrating on these common qualities improves how we might interpret the rationed components that oversee life.

5.2 Infection Models and Hereditary Issues

Angelfish, as living creatures with a hereditary outline, can act as models for grasping hereditary problems. While the points of interest of angelfish illnesses might contrast from human hereditary problems, the standards of hereditary qualities and the effect of changes on wellbeing are shared.

Researching hereditary issues in angelfish gives a stage to investigate possible helpful mediations and extend how we might interpret sickness systems.

6. **Natural Genomics: Adjusting to Changing Biological systems**

The investigation of angelfish hereditary qualities stretches out past the limits of aquariums and labs to the powerful environments they occupy. Ecological genomics explores how angelfish populaces answer changes in their normal environments, giving important bits of knowledge into the versatile procedures encoded in their genomes.

6.1 Genomic Reactions to Environmental Change

As environmental change adjusts amphibian biological systems, understanding how angelfish genomes answer becomes basic. Natural genomics dives into the genomic transformations that permit angelfish to adapt to temperature variances, living space debasement, and other ecological stressors. This information illuminates preservation rehearses pointed toward safeguarding the versatility of angelfish populaces despite an evolving environment.

6.2 Anthropogenic Effect on Genomic Variety

Human exercises, for example, overfishing, environment obliteration, and contamination, can apply specific tensions on angelfish populaces. Natural genomics disentangles the genomic results of anthropogenic effect, revealing insight into how human activities impact the hereditary variety and developmental directions of angelfish species. Protection methodologies informed by natural genomics plan to moderate the inconvenient impacts of human exercises on sea-going environments.

7. **Future Skylines: Coordinating Angelfish Hereditary qualities into Logical Exchanges**

As the investigation of angelfish hereditary qualities unfurls, what's in store holds promising skylines that reverberate a long ways past aquariums and research labs. Coordinating angelfish hereditary qualities into more extensive logical discoursed, including developmental science, environment, and bio-medical examination, upgrades our aggregate comprehension of the unpredictable associations inside the trap of life.

7.1 Transformative Exchanges: Disentangling Shared Narratives

The transformative exchanges among angelfish and different creatures, including people, keep on unwinding shared accounts written in the language of qualities. Near genomics turns into a device for interpreting the old connections and unique ways that have molded the biodiversity we notice today. By investigating the transformative discoursed implanted in angelfish hereditary qualities, we gain viewpoints on the interconnectedness of life on The planet.

7.2 Biomedical Experiences: Applications Past the Amphibian Domain

The utilizations of angelfish hereditary qualities stretch out past the limits of their watery natural surroundings. Biomedical experiences acquired from the investigation of angelfish hereditary qualities add to progressions in regenerative medication, immunology, and the comprehension of hereditary elements affecting sicknesses. The hereditary equals among angelfish and people become spans that interface the amphibian domain to the wildernesses of clinical examination.

8. **Moral Contemplations: Exploring the Convergence of Science and Preservation**

As the investigation of angelfish hereditary qualities propels, moral contemplations come to the very front. Dependable practices in logical examination, protection endeavors, and the aquarium exchange guarantee that the information acquired from concentrating on angelfish hereditary qualities lines up with moral standards and contributes emphatically to the prosperity of both the organic entities and the environments they occupy.

8.1 Moral Components of Hereditary Exploration on Sea-going Life forms

Hereditary examination on amphibian creatures, including angelfish, raises moral contemplations connected with the assortment of examples, lab trial and error, and the dispersal of hereditary data. Guaranteeing moral norms in hereditary examination includes a pledge to empathetic treatment, straightforwardness in trial methods, and the mindful sharing of discoveries inside mainstream researchers.

8.2 Moral Aquarium Practices and Protection Mindfulness

The aquarium exchange, driven by the appeal of elaborate fish like angelfish, requires moral contemplations to guarantee the supportability of wild populaces. Protection mindfulness crusades, moral reproducing practices, and endeavors to lessen the effect of the aquarium exchange on normal environments become essential parts of dependable aquarium the executives. Offsetting the interest with angelfish in imprisonment with the preservation of their wild partners requires a sensitive moral dance.

2.2 Comparative analysis of genomic findings between angelfish and humans.

The similar examination of genomic discoveries among angelfish and people offers a significant investigation into the common hereditary scenes that interface dissimilar parts of the tree of life. As the hereditary codes of these two species are translated, researchers uncover the consistent ideas that tight spot them as well as the disparate ways that have molded their one of a kind variations.

This relative genomics attempt gives bits of knowledge into the developmental cycles that have shaped the genomes of angelfish and people, revealing insight into the old hereditary exchanges that reverberate across different biological systems.

1. **Transformative Flexibility and Protection of Hereditary Components:**
   One of the striking disclosures from the near examination lies in the developmental versatility and protection of specific hereditary components. Divided qualities and administrative groupings among angelfish and people feature the antiquated starting points of essential natural cycles. Components liable for early stage advancement, resistant reaction, and metabolic pathways show striking preservation, highlighting the crucial jobs these cycles play in the endurance of different species. The conservation of these hereditary components across developmental time addresses their essentialness in the excellent ensemble of life.

2. **Bits of knowledge into Infection Models and Hereditary Issues:**
   Similar genomics among angelfish and people offers a one of a kind stage for investigating sickness models and hereditary problems. While the particular sicknesses might vary, the standards of hereditary qualities and the effect of changes on wellbeing stay widespread. Examining hereditary problems in angelfish gives a significant viewpoint, permitting researchers to concentrate on the indication of changes, their consequences for physiological cycles, and likely restorative mediations. This cross-species approach not just improves how we might interpret explicit hereditary issues yet additionally lays the preparation for translational exploration with suggestions for human wellbeing.

3. **Hereditary Varieties and Transformations:**

The similar investigation dives into the hereditary varieties that add to the transformations of angelfish and people to their separate surroundings. The two species have gone through transformative cycles that have formed their hereditary cosmetics in light of ecological difficulties. Concentrating on these varieties gives experiences into the versatile techniques encoded in their genomes, from blade morphology and variety designs in angelfish to physiological reactions and illness opposition in people. Understanding the hereditary premise of variations adds to a nuanced enthusiasm for the powerful interaction among organic entities and their biological systems.

4. **Protection Suggestions and Biodiversity Conservation:**

   The genomic matches among angelfish and people hold suggestions for protection science. The protection of specific hereditary components accentuates the interconnectedness of biological systems and the significance of safeguarding biodiversity.

   Similar genomics supports recognizing versatility qualities related with natural transformation, offering significant bits of knowledge for protection systems. By disentangling the hereditary underpinnings of variation and variety, researchers can add to the safeguarding of angelfish populaces in their normal environments, encouraging a comprehensive way to deal with marine protection.

5. **Translational Open doors in Biomedical Exploration:**

The near examination opens roads for translational open doors in biomedical exploration. Hereditary experiences acquired from angelfish can possibly illuminate headways in regenerative medication, immunology, and the comprehension of hereditary elements impacting sicknesses. The common hereditary standards become spans that interface the oceanic domain to the boondocks of clinical exploration. As researchers explore the translational scene, the similar genomics approach guarantees creative applications with suggestions for human wellbeing.

2.3 Potential applications in understanding genetic disorders.

The investigation of angelfish hereditary qualities presents an extraordinary chance for disentangling the secrets of hereditary problems, offering potential applications that stretch out past the oceanic domain. While angelfish may not straightforwardly share the range of hereditary issues with people, the essential standards of hereditary qualities and the effect of changes on wellbeing are moderated across species. Here are possible applications in figuring out hereditary problems through the investigation of angelfish hereditary qualities:

1. **Illness Models and Systems:**
   Angelfish act as important models for concentrating on the appearance of hereditary transformations and their consequences for physiological cycles.

By prompting explicit hereditary changes, scientists can notice the advancement of qualities and conditions similar to hereditary issues. This empowers a more profound comprehension of the fundamental instruments, assisting with clarifying how explicit qualities add to the indication of sicknesses. The experiences acquired from angelfish models give a scaffold between essential hereditary standards and the complexities of infection movement.

2. **Translational Exploration for Human Wellbeing:**

The translational capability of angelfish hereditary qualities lies in its capacity to illuminate research with suggestions for human wellbeing. Revelations made in angelfish models can be made an interpretation of to acquire bits of knowledge into comparative hereditary pathways and components in people. This cross-species approach upgrades how we might interpret rationed hereditary components, making ready for imaginative restorative intercessions and medication advancement. The information acquired from angelfish hereditary qualities turns into an asset for translational exploration, offering an exceptional point of view on the hereditary premise of illnesses.

3. **Recognizable proof of Helpful Targets:**

Concentrating on angelfish hereditary qualities adds to the ID of likely helpful focuses for hereditary problems. As scientists uncover the qualities related with explicit attributes or conditions in angelfish, these discoveries might resemble human hereditary issues. Focusing on these saved hereditary components gives an establishment to creating helpful systems that can tweak quality articulation, right transformations, or relieve the impacts of hereditary issues. The distinguishing proof of restorative focuses through angelfish hereditary qualities lines up with the more extensive objective of accuracy medication and customized treatment draws near.

4. **Experiences into Hereditary Variety and Illness Vulnerability:**

The investigation of angelfish hereditary qualities reveals insight into hereditary varieties and their effect on illness vulnerability. Understanding how hereditary varieties add to transformations and reactions to ecological stressors in angelfish gives experiences into comparable cycles in people. This similar investigation of hereditary variety offers a more extensive viewpoint on the interaction between hereditary variety and infection powerlessness, adding to a more nuanced comprehension of the hereditary variables impacting wellbeing and sickness.

Generally, the likely utilizations of angelfish hereditary qualities in understanding hereditary issues stretch out from filling in as illness models to offering translational bits of knowledge for human wellbeing. The saved hereditary standards divided among angelfish and people set out an exceptional freedom to unwind the intricacies of hereditary issues, making ready for headways in conclusion, treatment, and our more extensive comprehension of the hereditary underpinnings of wellbeing and illness.

Chapter 3

Cardiovascular Discoveries

The cardiovascular framework, with its complicated organization of conduits, veins, and the throbbing heart at its center, has for some time been a point of convergence of logical request. From the early physical investigations of the circulatory framework to current sub-atomic examinations, cardiovascular revelations play had a critical impact in propelling comprehension we might interpret heart wellbeing, illness systems, and restorative mediations. This far reaching investigation traverses the verifiable tourist spots, flow forward leaps, and future outskirts in cardiovascular examination, digging into the intricacies that characterize the pulsating heart of life.

1. **Authentic Points of view: Physical Establishments and Early Bits of knowledge**

   1.1 Early Physical Investigations:

   The underlying foundations of cardiovascular revelations follow back to old civic establishments, where early anatomists tried to disentangle the secrets of the human body. The antiquated Egyptians, Greeks, and Romans made huge commitments to figuring out the circulatory framework, establishing the groundwork for future examinations.

   1.2 Harvey's Progressive Circulatory Hypothesis:

   The seventeenth century saw a change in perspective with the notable work of William Harvey. His hypothesis of blood course, introduced in "De Motu Cordis," tested winning convictions and presented the idea of a shut circulatory framework. Harvey's disclosures laid the preparation for future examinations concerning the elements of blood stream and cardiovascular capability.

2. **Present day Bits of knowledge into Cardiovascular Construction and Capability**

2.1 Advances in Imaging Advancements:

The twentieth century carried extraordinary changes to cardiovascular exploration, set apart by the appearance of cutting edge imaging advancements. From X-beams to echocardiography and attractive reverberation imaging (X-ray), these apparatuses permitted analysts to envision the unpredictable designs of the heart, analyze conditions, and screen cardiovascular capability with extraordinary detail.

2.2 Electrocardiography and Musicality Issues:

The advancement of electrocardiography by Willem Einthoven in the mid twentieth century altered the determination of heart cadence problems. The electrocardiogram (ECG) turned into a foundation in grasping the electrical action of the heart, prompting experiences into arrhythmias, conduction irregularities, and the distinguishing proof of basic cardiovascular circumstances.

3. Sub-atomic Systems: Unwinding the Hereditary Embroidered artwork of Cardiovascular Wellbeing

3.1 Genomic Experiences into Cardiovascular Sicknesses:

The genomic period introduced another outskirts in cardiovascular examination, permitting researchers to investigate the hereditary premise of cardiovascular sicknesses. Vast affiliation studies (GWAS) distinguished hereditary variations related with conditions like hypertension, coronary conduit illness, and cardiovascular breakdown, giving hints to the fundamental components and expected helpful targets.

3.2 Epigenetics and Cardiovascular Wellbeing:

Epigenetic changes, including DNA methylation and histone acetylation, arose as significant players in cardiovascular wellbeing. The powerful transaction among hereditary and epigenetic factors impacts quality articulation, adding to the turn of events and movement of cardiovascular illnesses. Understanding these epigenetic components opens roads for designated mediations and customized medication draws near.

4. Cardiovascular Illnesses: From Atherosclerosis to Cardiovascular breakdown

4.1 Atherosclerosis and Coronary Vein Sickness:

Atherosclerosis, described by the gathering of plaques in blood vessel walls, stays a main source of cardiovascular grimness and mortality. Propels in understanding the sub-atomic pathways engaged with plaque arrangement, burst, and apoplexy have prompted the improvement of imaginative treatments, including statins, antiplatelet specialists, and arising mediations focusing on fiery pathways.

4.2 Cardiovascular breakdown and Remedial Developments:

Cardiovascular breakdown, a perplexing condition with disabled heart capability, presents critical difficulties in cardiovascular medication. Research has revealed atomic pathways engaged with heart rebuilding, neurohormonal initiation, and contractile brokenness. Novel treatments, for example, angiotensin-changing over compound (ACE) inhibitors, beta-blockers, and gadgets like implantable cardioverter-defibrillators (ICDs), have changed the administration of cardiovascular breakdown.

5. **Interventional Cardiology: Altering Treatment Approaches**

5.1 Percutaneous Coronary Mediations:

The approach of percutaneous coronary mediations (PCIs), including angioplasty and stent position, altered the treatment of coronary vein illness. These negligibly obtrusive methods plan to reestablish blood stream to the heart, mitigate side effects, and further develop results in patients with ischemic coronary illness.

5.2 Transcatheter Valve Treatments:

Advancements in transcatheter valve treatments have given elective choices to patients with valvular coronary illness. Transcatheter aortic valve substitution (TAVR) and transcatheter mitral valve fix address primary heart issues without the requirement for open-heart medical procedure, offering additional opportunities for high-chance and older patients.

6. **Regenerative Medication and Immature microorganism Treatments**

6.1 Heart Recovery and Fix:

Regenerative medication holds guarantee in tending to the test of heart tissue fix. Analysts investigate ways to deal with animate the recovery of harmed heart tissue, tackling the capability of undeveloped cells, quality treatment, and tissue designing. The objective is to reestablish practical myocardium and work on cardiovascular capability after wounds like myocardial dead tissue.

6.2 Undifferentiated cell Treatments for Coronary illness:

Foundational microorganism treatments, including the utilization of mesenchymal undifferentiated organisms and actuated pluripotent undeveloped cells, have been examined for their regenerative potential in coronary illness. Clinical preliminaries mean to assess the security and viability of these treatments in upgrading heart fix and capability, proclaiming another time in cardiovascular regenerative medication.

7. **Accuracy Medication and Individualized Cardiovascular Consideration**

7.1 Pharmacogenomics in Cardiovascular Medication:

The reconciliation of pharmacogenomics into cardiovascular medication addresses a step towards accuracy medication. Understanding what a person's hereditary cosmetics means for drug reaction considers customized treatment draws near, upgrading adequacy and limiting unfavorable impacts in conditions like hypertension, dyslipidemia, and thrombotic messes.

7.2 Omics Innovations and Biomarker Disclosure:

Omics innovations, including genomics, proteomics, and metabolomics, add to biomarker disclosure for cardiovascular illnesses. ID of explicit sub-atomic marks empowers early location, risk definition, and checking of illness movement. Biomarker-driven approaches improve analytic exactness and work with designated mediations in customized cardiovascular consideration.

8. **Cardiovascular Wellbeing Advancement and Populace Intercessions**

    8.1 Way of life Intercessions and Anticipation:

Cardiovascular wellbeing advancement underscores way of life mediations as essential parts in forestalling coronary illness. Dietary adjustments, customary active work, smoking suspension, and stress the board assume vital parts in decreasing cardiovascular gamble factors. General wellbeing efforts highlight the significance of way of life decisions in keeping up with heart wellbeing across populaces.

8.2 Worldwide Wellbeing Drives and Admittance to Cardiovascular Consideration:

Worldwide wellbeing drives expect to address differences in cardiovascular consideration and further develop admittance to fundamental mediations around the world. Endeavors to execute practical techniques, upgrade medical services framework, and give schooling on cardiovascular wellbeing add to decreasing the worldwide weight of coronary illness.

9. **Arising Advancements and Future Wildernesses in Cardiovascular Exploration**

    9.1 Computerized reasoning and Prescient Examination:

Computerized reasoning (artificial intelligence) and AI applications in cardiovascular exploration can possibly change diagnostics, risk expectation, and treatment arranging. Prescient examination influence enormous datasets to recognize designs, empowering early ID of people in danger for cardiovascular sicknesses and directing customized mediations.

9.2 CRISPR Quality Altering in Cardiovascular Exploration:

The CRISPR-Cas9 quality altering apparatus opens new outskirts in cardiovascular exploration by permitting exact adjustments to the hereditary code. Analysts investigate the use of CRISPR innovation to explore hereditary variables basic cardiovascular infections, concentrate on sickness systems, and foster likely restorative intercessions.

10. **Moral Contemplations and Cultural Ramifications**

10.1 Moral Contemplations in Cardiovascular Exploration:

The progression of cardiovascular exploration prompts moral contemplations connected with patient assent, information protection, and the capable direct of clinical preliminaries. Guaranteeing straightforwardness, regarding independence, and protecting patient freedoms are vital in exploring the moral scene of cardiovascular examination.

10.2 Cultural Ramifications of Cardiovascular Revelations:

Cardiovascular revelations have significant cultural ramifications, affecting medical services approaches, asset designation, and public mindfulness. The weight of cardiovascular illnesses on medical services frameworks highlights the significance of preventive systems, early mediation, and ceaseless exploration endeavors to address the advancing difficulties presented by heart-related conditions.

3.1Examination of angelfish cardiovascular systems and similarities to human hearts.

The investigation of cardiovascular frameworks in angelfish uncovers a captivating section in near science, drawing equals and qualifications between the sea-going domain and the mind boggling heart hardware of people. This thorough assessment envelops physical designs, physiological cycles, and developmental variations, revealing insight into the common rules that oversee the cadenced dance of hearts across species.

1. **Life systems of the Angelfish Heart: A Watery Orchestra**

    1.1 Cardiovascular Chambers and Construction:

    The angelfish heart, a wonderful organ inside its sea-going living space, shows a two-chambered structure including the chamber and ventricle. The chamber gets deoxygenated blood, while the ventricle moves oxygenated blood all through the circulatory framework. This improved on cardiovascular plan stands out from the more mind boggling four-chambered heart of people however highlights the proficiency of angelfish in their sea-going conditions.

    1.2 Valves and Circulatory Pathways:

    Valves inside the angelfish heart assume a vital part in coordinating blood stream and forestalling discharge. Analyzing the usefulness of these valves uncovers transformations customized to the hydrodynamic difficulties of submerged life. The circulatory pathways, however smoothed out, exhibit accuracy in guiding blood to fundamental organs and tissues, displaying the transformative variations that advance cardiovascular proficiency in angelfish.

2. **Physiological Cycles: Cadenced Stories of Hearts**

    2.1 Pulse and Circulatory Requests:

    The assessment of angelfish physiology digs into the complexities of pulse guideline and circulatory requests. These fish, possessing different amphibian conditions, show versatility in changing their pulses to ecological boosts. The physiological reactions reflect the unique cooperation among angelfish and their submerged environments, featuring the job of the cardiovascular framework in gathering the metabolic necessities of assorted territories.

    2.2 Oxygen Take-up and Vaporous Trade:

Vaporous trade in angelfish happens essentially through gills, underlining the significant connection between the cardiovascular and respiratory frameworks. The assessment of oxygen take-up components uncovers the effectiveness with which angelfish extricate oxygen from water, supporting vigorous digestion and supporting their dynamic ways of life. Correlations with human respiratory and cardiovascular variations offer experiences into the developmental pathways molded by particular ecological tensions.

3. **Transformative Variations: Exploring Sea-going Domains**

   3.1 Hydrodynamic Difficulties and Cardiovascular Plan:

   The oceanic way of life of angelfish forces remarkable hydrodynamic difficulties that have shaped their cardiovascular plan. The smoothed out bodies and productive hearts reflect transformations improved for swimming proficiency, with cardiovascular frameworks finely tuned to explore flows and differing water conditions. Looking at these variations gives a window into the transformative discoursed that have formed the cardiovascular flexibility of angelfish.

   3.2 Lightness, Blood Sythesis, and Osmoregulation:

   Lightness, a basic calculate the sea-going domain, entwines with the angelfish cardiovascular framework's job in keeping up with suitable blood sythesis. Osmoregulation, a mind boggling dance among water and particles, impacts cardiovascular transformations. Figuring out the fragile equilibrium in blood structure and osmoregulatory components improves the enthusiasm for how angelfish hearts add to the general wellness of these oceanic organic entities.

4. **Similar Investigation with Human Hearts: Equals and Divergences**

   4.1 Chambered Hearts and Circulatory Effectiveness:

   Contrasting angelfish hearts with their human partners uncovers the two equals and divergences. While angelfish have a two-chambered heart improved for sea-going life, people gloat a more mind boggling four-chambered heart that upholds the requests of earthly presence.

   Investigating these primary distinctions gives bits of knowledge into the development of heart chambers and the compromises between circulatory productivity and ecological specialization.

   4.2 Valvular Usefulness and Hemodynamics:

   The assessment of valves in angelfish hearts, intended for ideal submerged blood stream, appears differently in relation to the multifaceted valvular designs in human hearts. Contrasts in valvular usefulness impact hemodynamics, with suggestions for the energetics of blood course. Understanding these varieties adds to a nuanced enthusiasm for the heart transformations that have developed in light of unmistakable natural specialties.

5. **Shared Standards: The Embodiment of Cardiovascular All inclusiveness**

5.1 Heart Compression and Muscle Physiology:

At the center of both angelfish and human hearts lies the basic guideline of cardiovascular withdrawal. Analyzing the similitudes in muscle physiology, contractile proteins, and the coordination of musical beats enlightens the general substance of cardiovascular capability. The common standards highlight the old hereditary discoursed that interface assorted species through the language of the heart.

5.2 Blood Flow and Foundational Coordination:

The assessment of blood flow designs in angelfish and people uncovers foundational reconciliation as an ongoing idea. In spite of primary divergences, the general objective of providing oxygen and supplements to crucial tissues stays consistent. Investigating the common standards of blood flow accentuates the versatile idea of cardiovascular frameworks across transformative scenes.

6. **Future Headings in Near Cardiovascular Exploration**

6.1 Incorporating Genomic Bits of knowledge:

Future bearings in relative cardiovascular exploration include coordinating genomic bits of knowledge to disentangle the hereditary underpinnings of cardiovascular transformations. Similar genomics holds the way to translating the rationed and different hereditary components that shape the cardiovascular frameworks of angelfish and people. Understanding the hereditary exchanges implanted in heart genomes vows to divulge the sub-atomic underpinnings of cardiovascular flexibility and specialization.

6.2 Ramifications for Human Wellbeing:

The assessment of angelfish cardiovascular frameworks improves how we might interpret oceanic variations as well as conveys ramifications for human wellbeing.

Translational applications, roused by transformative standards, may illuminate cardiovascular treatments, biomimetic plans, and advancements in regenerative medication. Crossing over the information acquired from angelfish hearts to headways in human cardiovascular wellbeing highlights the interconnectedness of natural frameworks.

7. **Moral Contemplations and Protection Suggestions**

7.1 Moral Components of Relative Exploration:

As specialists dive into the complexities of angelfish cardiovascular frameworks, moral contemplations guide the mindful direct of relative examination. Guaranteeing altruistic treatment, straightforwardness in exploratory systems, and aware commitment with living life forms line up with moral norms in natural examinations. Moral aspects highlight the significance of moral appearance chasing after information.

7.2 Protection Suggestions for Amphibian Biological systems:

The assessment of angelfish cardiovascular frameworks conveys suggestions for the preservation of sea-going biological systems. Understanding the cardiovascular variations that empower these fish to flourish in assorted conditions illuminates protection systems. Offsetting logical investigation with biological stewardship becomes principal, stressing the interconnectedness of cardiovascular examination and the prosperity of oceanic environments.

3.2 Insights into heart development and potential implications for cardiac health.

The excursion of life starts with the musical dance of a little heart, a wonder of natural designing that develops from a fragile undeveloped construction into the mind boggling organ that supports our reality. Investigating the experiences into heart improvement divulges the perplexing movement of cell processes, hereditary arrangements, and ecological impacts that shape the diagram of life. Past the domains of embryology, these experiences have significant ramifications for grasping cardiovascular wellbeing, molding helpful mediations, and preparing for headways in regenerative medication.

1. **Undeveloped Heart Advancement: The Ensemble Starts**

   1.1 Development of the Crude Heart Cylinder:

   Undeveloped heart improvement initiates with the development of the crude heart tube, a transient construction that establishes the groundwork for the complex cardiovascular design. Exact flagging pathways organize the relocation, multiplication, and separation of cells, changing a straightforward rounded structure into the throbbing focus of life. Understanding the sub-atomic signals that guide these beginning phases offers bits of knowledge into the major cycles that shape cardiovascular predetermination.

   1.2 Cardiomyocyte Separation and Chamber Arrangement:

   The separation of cardiomyocytes, the particular muscle cells of the heart, denotes a vital stage in undeveloped turn of events. Unmistakable populaces of cells add to the development of atrial and ventricular chambers, each with a novel job in coordinating the musical compressions that move blood through the circulatory framework. Analyzing the sub-atomic signs and hereditary controllers engaged with cardiomyocyte separation disentangles the complexities of chamber arrangement.

2. **Hereditary Establishments: Deciphering the Heart's Hereditary Orchestra**

   2.1 Job of Record Elements in Cardiovascular Destiny Assurance:

   Record factors arise as maestros in the hereditary orchestra of heart improvement, coordinating the destiny of forerunner cells toward a cardiovascular predetermination. Ace controllers, for example, Nkx2.5 and GATA4, guide the transcriptional scene, actuating qualities fundamental for cardiomyocyte improvement. Digging into the jobs of these record factors gives key bits of knowledge into the hereditary establishments that underlie the organization of heart improvement.

2.2 Flagging Pathways and Morphogen Slopes:

Signal transduction pathways, including Wnt, BMP, and Score, lay out morphogen inclinations that direct cell choices during heart advancement. Spatial and worldly signs shape the situating of heart structures, guaranteeing the exact arrangement of chambers and valves. Translating the language of flagging pathways enlightens the cross-talk among cells and their current circumstance, uncovering the complicated dance that shapes the creating heart.

3. **Cell Elements: Chiseling the Engineering Work of art**

3.1 Cardiovascular Morphogenesis and Circling:

Cardiovascular morphogenesis changes the direct heart tube into a circled structure, a basic step that lays out the left-right lopsidedness of the heart. The organized developments of cells during circling decide the spatial plan of cardiovascular chambers and the arrangement of the blood vessel and venous posts. Disentangling the cell elements of circling gives a window into the design chiseling of the creating heart.

3.2 Endocardial Pad Arrangement and Valve Improvement:

Endocardial pads, particular districts inside the heart tube, assume a significant part in valve improvement. Mesenchymal cells got from these pads add to the arrangement of atrioventricular and semilunar valves, critical parts that control blood stream. Researching the cell occasions basic endocardial pad arrangement reveals insight into the components that guarantee the underlying respectability of the creating heart.

4. **Ecological Impacts: Organizing the Heart's Reaction**

4.1 Hemodynamic Powers and Heart Improvement:

Hemodynamic powers, produced by the progression of blood inside the creating heart, apply significant impacts on cardiovascular morphogenesis. Shear pressure and tension inclinations shape the underlying variations of heart tissues, impacting cell multiplication, valve arrangement, and chamber rebuilding. Inspecting the interchange between hemodynamic powers and heart advancement uncovers the unique responsiveness of the creating heart to its mechanical climate.

4.2 Maternal Variables and Epigenetic Impacts:

Past the limits of the incipient organism, maternal variables and natural prompts add to the arrangement of heart advancement. Epigenetic changes, including DNA methylation and histone acetylation, intervene the intergenerational transmission of data that impacts heart aggregates. Understanding the natural and epigenetic effects on heart advancement expands the extent of experiences into the elements forming cardiovascular wellbeing.

5. **Suggestions for Cardiovascular Wellbeing: Crossing over Improvement and Infection**

5.1 Formative Beginnings of Heart Sicknesses:

Experiences into heart improvement disclose the formative starting points of heart illnesses, underscoring the meaning of occasions during undeveloped life in inclining people toward cardiovascular circumstances. Distortions in early cardiovascular morphogenesis, hereditary controllers, or ecological impacts can make way for inherent heart surrenders, valvular anomalies, and other primary issues. Overcoming any issues among improvement and sickness gives a comprehensive point of view on the continuum of cardiovascular wellbeing.

5.2 Intrinsic Heart Imperfections: From Morphogenesis to Clinical Indications:

The comprehension of heart improvement advises the cognizance regarding intrinsic heart surrenders, going from septal imperfections to anomalies in chamber arrangement. Sub-atomic bits of knowledge into the hereditary premise of intrinsic heart sicknesses enlighten the pathways disturbed in these circumstances. Connecting formative components with clinical signs improves demonstrative accuracy, restorative mediations, and the advising of families impacted by inborn heart surrenders.

6. **Helpful Ramifications: Directing Developments in Cardiovascular Consideration**

6.1 Regenerative Medication and Cardiovascular Fix:

Bits of knowledge into heart improvement fuel developments in regenerative medication, offering expect cardiovascular fix and recovery. Systems including undifferentiated organisms, tissue designing, and quality treatments plan to tackle the regenerative possible innate in formative cycles. Making an interpretation of formative bits of knowledge into helpful mediations proclaims another time in the journey for regenerative ways to deal with treat cardiovascular illnesses.

6.2 Accuracy Medication in Cardiology:

The assembly of formative experiences with accuracy medication makes ready for fitted ways to deal with heart care. Understanding the hereditary and ecological determinants that shape individual cardiovascular wellbeing empowers customized risk evaluations, early mediations, and designated treatments. Accuracy medication in cardiology use the mind boggling information acquired from heart advancement to streamline patient results and upgrade the nature of cardiovascular medical care.

7. **Future Wildernesses: Exploring Neglected Domains**

7.1 Advances in Formative Science and Innovation:

The eventual fate of heart improvement research unfurls against a scenery of propelling advancements and systems in formative science. State of the art strategies, including single-cell RNA sequencing, CRISPR quality altering, and organoids, give extraordinary bits of knowledge into the sub-atomic subtleties of heart advancement. These innovative wildernesses vow to unwind already unavailable intricacies, directing analysts into unfamiliar regions of formative science.

7.2 Combination of Multi-Omics Approaches:

The combination of multi-omics approaches, enveloping genomics, transcriptomics, proteomics, and metabolomics, holds the way in to a complete comprehension of heart improvement. Unwinding the sub-atomic marks that characterize each phase of cardiovascular morphogenesis gives a nuanced enthusiasm for the unpredictable administrative organizations overseeing the development of a useful heart. Multi-omics combination upgrades the profundity and expansiveness of experiences into the organization of life's most memorable heartbeat.

8. **Moral Contemplations: Supporting Mindful Exploration**

8.1 Moral Aspects in Formative Exploration:

The investigation of heart improvement requires a promise to moral rules that defend the government assistance of examination subjects and maintain the uprightness of logical request. Guaranteeing straightforwardness, informed assent, and aware treatment of living creatures become foremost in exploring the moral elements of formative examination. Moral contemplations guide the capable direct of tests and the scattering of information chasing a more profound comprehension of life's starting points.

3.3Implications for cardiovascular disease research.

The complicated bits of knowledge acquired from the investigation of heart improvement stretch out a long ways past the domains of embryology, venturing into the center of cardiovascular infection research. As we interpret the sub-atomic exchanges, hereditary organizations, and ecological impacts that shape the heart's outline, we uncover a mother lode of information with significant ramifications for understanding, treating, and forestalling cardiovascular infections. This far reaching investigation dives into the ramifications for cardiovascular infection research, enlightening the way toward helpful developments, accuracy medication, and an all encompassing way to deal with heart wellbeing.

1. **Unwinding the Sub-atomic Premise of Cardiovascular Sicknesses**

   1.1 Hereditary Groundworks of Cardiovascular Illnesses:

The comprehension of heart improvement gives a nuanced viewpoint on the hereditary groundworks of cardiovascular sicknesses. Experiences into the expert controllers and flagging pathways that guide early stage cardiovascular morphogenesis uncover matches with the sub-atomic underpinnings of heart illnesses. The investigation of normal hereditary components turns into a reference point for specialists, directing them toward the ID of central members embroiled in the turn of events and movement of cardiovascular problems.

1.2 Connecting Improvement and Infection: The Continuum of Heart Wellbeing:

The continuum from heart improvement to cardiovascular illnesses addresses a scaffold that traverses the range of heart wellbeing. Abnormalities in formative cycles might make way for inborn heart deserts, while varieties in hereditary controllers might add to the helplessness to gained heart sicknesses further down the road. Perceiving the interconnectedness of improvement and illness guides specialists in uncovering shared components, offering a comprehensive system for cardiovascular sickness research.

2. **Innate Heart Deformities: From Morphogenesis to Restorative Bits of knowledge**

2.1 Early Finding and Intercession:

Bits of knowledge from heart improvement upgrade our capacity to analyze and mediate in inborn heart absconds. Understanding the formative starting points of these circumstances empowers early ID of people in danger, considering convenient mediations and customized treatment plans. The utilization of hereditary information to inherent heart illnesses cultivates a change in perspective toward accuracy medication, where helpful procedures are customized to the exceptional hereditary and formative profiles of impacted people.

2.2 Improvement Roused Treatments:

The equals between heart improvement and innate heart abandons motivate creative restorative methodologies. Drawing from the regenerative possible inborn in early stage processes, specialists investigate procedures for heart fix and recovery. Undifferentiated cell treatments, tissue designing, and quality altering advances hold guarantee in rectifying primary irregularities and reestablishing typical cardiovascular capability. The reconciliation of formative bits of knowledge into remedial systems denotes a groundbreaking time in inherent heart deformity research.

3. **Procured Cardiovascular Sicknesses: Unraveling the Illness Scene**

3.1 Atherosclerosis and Coronary Conduit Sickness:

The atomic experiences accumulated from heart improvement shed light on the pathogenesis of obtained cardiovascular sicknesses, like atherosclerosis and coronary supply route illness. The comprehension of hereditary controllers, flagging pathways, and cell elements becomes urgent in unwinding the components driving the arrangement of atherosclerotic plaques. This more profound understanding aides analysts in creating designated mediations pointed toward forestalling, ending, or turning around the movement of these common cardiovascular illnesses.

3.2 Valve Problems and Primary Irregularities:

Valvular problems and primary irregularities, both inborn and obtained, track down associations with the perplexing cell cycles of heart improvement. Disentangling the formative starting points of valves gives a guide to examining oddities in valve design and capability. Research endeavors directed by formative bits of knowledge improve demonstrative exactness, illuminate careful mediations, and prepare for imaginative ways to deal with fix or supplant harmed valves.

4. **Crossing over Hereditary qualities and Accuracy Medication in Cardiovascular Consideration**

4.1 Hereditary Screening and Hazard Delineation:

The combination of hereditary information got from heart improvement into cardiovascular sickness research proclaims another period in hereditary screening and hazard separation. Distinguishing hereditary variations related with an expanded gamble of cardiovascular infections empowers proactive measures for people with increased vulnerability. Early intercessions, way of life changes, and designated treatments become fundamental parts of an accuracy medication approach that is custom-made to individual hereditary profiles.

4.2 Pharmacogenomics and Customized Treatment:

The field of pharmacogenomics, enhanced by experiences from heart advancement, adds to customized treatment techniques in cardiovascular consideration. Understanding how individual hereditary varieties impact drug reaction considers the customization of medicine regimens. Custom fitted medication treatments in light of hereditary profiles upgrade treatment adequacy, limit antagonistic impacts, and streamline cardiovascular results. The marriage of hereditary qualities and pharmacology turns into a foundation in the developing scene of accuracy cardiovascular medication.

5. **Regenerative Medication and Novel Restorative Boondocks**

5.1 Tackling Formative Regenerative Potential:

The regenerative potential inborn in heart improvement turns into a wellspring of motivation for novel remedial outskirts. Analysts investigate methodologies to saddle the regenerative limits saw during undeveloped heart morphogenesis. Undifferentiated cell treatments, tissue designing methodologies, and quality altering advances expect to invigorate heart fix and recovery, offering expect people impacted via cardiovascular illnesses. The interpretation of formative bits of knowledge into regenerative medication opens roads for groundbreaking treatments with the possibility to alter cardiovascular consideration.

5.2 Quality Treatments and CRISPR-Cas9:

Progressions in quality treatments and the progressive CRISPR-Cas9 quality altering device present phenomenal open doors in cardiovascular sickness research. Understanding the hereditary underpinnings of heart advancement permits specialists to target explicit qualities embroiled in cardiovascular illnesses. Accuracy quality altering procedures hold guarantee for amending hereditary anomalies, balancing sickness related pathways, and propelling helpful mediations with the possibility to address the main drivers of cardiovascular problems.

6. **Comprehensive Ways to deal with Cardiovascular Wellbeing**

   6.1 Way of life Intercessions Informed by Formative Bits of knowledge:

The all encompassing ramifications of formative bits of knowledge reach out to way of life intercessions pointed toward advancing cardiovascular wellbeing. Perceiving the formative beginnings of heart infections highlights the significance of early-life impacts on cardiovascular results. General wellbeing drives, instructive missions, and intercessions that advance sound ways of life draw motivation from formative information, cultivating a proactive way to deal with cardiovascular wellbeing from youth through adulthood.

   6.2 Integrative Cardiovascular Consideration:

An integrative way to deal with cardiovascular consideration arises as a characteristic outcome of figuring out the formative continuum. Crossing over the universes of intrinsic heart surrenders, procured cardiovascular infections, and regenerative medication, integrative consideration embraces an extensive comprehension of the variables impacting heart wellbeing. Coordinated effort between cardiologists, geneticists, formative scientists, and regenerative medication experts becomes fundamental in giving all encompassing consideration that tends to the different components of cardiovascular wellbeing.

7. **Translational Exploration and Seat to-Bedside Applications**

   7.1 Making an interpretation of Formative Bits of knowledge to Clinical Applications:

The translational capability of formative bits of knowledge holds the way to seat to-bedside applications in cardiovascular sickness research. Deciphering information acquired from heart improvement into clinical applications upgrades demonstrative precision, illuminates restorative methodologies, and changes the scene of cardiovascular consideration. Cooperative endeavors between essential researchers and clinicians overcome any barrier between crucial exploration and its genuine effect on quiet results.

   7.2 From Exploration Revelations to General Wellbeing Effect:

The excursion from research revelations to general wellbeing influence exemplifies a definitive objective of cardiovascular sickness research. The incorporation of formative experiences into general wellbeing drives, preventive measures, and helpful advancements can possibly altogether diminish the weight of cardiovascular infections on worldwide wellbeing. Changing exploration revelations into unmistakable advantages for people and populaces underscores the cultural effect and general wellbeing pertinence of cardiovascular sickness research.

8. **Future Skylines and Moral Contemplations**

8.1 Arising Innovations and Future Skylines:

The future skylines of cardiovascular infection research unfurl against a scenery of arising innovations, creative techniques, and developing standards. Propels in single-cell omics, man-made brainpower, and organoid advances vow to open new elements of grasping in the unpredictable universe of cardiovascular wellbeing. The nonstop reconciliation of state of the art innovations into cardiovascular exploration impels the field toward unknown regions and novel leap forwards.

8.2 Moral Contemplations in Cardiovascular Sickness Exploration:

As the excursion through cardiovascular infection research advances, moral contemplations stay at the front. Guaranteeing mindful direct, straightforward correspondence, and the deferential treatment of exploration subjects, including those contributing hereditary data, becomes basic. The moral components of genomic research, regenerative medication, and restorative advancements request progressing watchfulness to maintain the standards of value, independence, and equity.

Chapter 4

Neurological Revelations

The human cerebrum, a wonder of organic intricacy, has long charmed the interest of researchers, scholars, and masterminds since forever ago. Late headways in neuroscience have impelled how we might interpret the mind to uncommon levels, uncovering an embroidery of multifaceted cell collaborations, hereditary codes, and dynamic brain organizations. This thorough investigation, crossing the domains of neuroanatomy, neurophysiology, and neurogenetics, dives into the neurological disclosures that have reshaped how we might interpret the human cerebrum and its significant ramifications for wellbeing, discernment, and the actual pith of human experience.

1. **Neuroanatomy: Exploring the Scene of the Cerebrum**

   1.1 Primary Wonders: The Structural Miracles of the Mind:

   The investigation of neuroanatomy uncovers the underlying wonders that characterize the scene of the cerebrum. From the tangled folds of the cerebral cortex to the complex hardware of the limbic framework, every locale adds to the arrangement of mental, profound, and tactile capabilities. The excursion through the mind's life systems disentangles the secrets of how its actual construction underlies the tremendous collection of human encounters.

   1.2 Connectomics: Planning the Brain Organizations:

   In the time of connectomics, specialists use progressed imaging procedures to plan the multifaceted brain networks that structure the underpinning of mind capability. The human cerebrum's network, both nearby and long-range, arises as a critical determinant of mental cycles. Unwinding the wiring graph of the mind upgrades how we might interpret data handling, memory arrangement, and the coordination of exercises across different cerebrum locales.

2. **Neurophysiology: The Orchestra of Brain Movement**

2.1 Neuronal Correspondence: Electrical Motivations and Compound Flagging:

At the core of neurophysiology lies the many-sided dance of neuronal correspondence, where electrical motivations and synthetic flagging organize the orchestra of brain movement. Neurons, the crucial units of the sensory system, send data through activity possibilities and synapses. The investigation of neurophysiological cycles uncovers the powerful interchange among excitation and hindrance that shapes brain circuits and underlies mental capabilities.

2.2 Synaptic Versatility: Adjusting to Experience:

The peculiarity of synaptic versatility arises as a foundation of neurophysiological disclosures. Neurotransmitters, the intersections between neurons, show dynamic changes in strength and network in light of involvement. Long haul potentiation and despondency, crucial systems of synaptic pliancy, assume urgent parts in learning, memory, and the versatile reactions of the mind to its consistently evolving climate.

3. **Neurogenetics: Interpreting the Hereditary Code of the Mind**

3.1 Genomic Experiences into Mental health:

Headways in neurogenetics give extraordinary experiences into the hereditary code that oversees mental health. Key formative qualities, like those associated with neurogenesis and brain movement, shape the early stage cerebrum's design. The transaction between hereditary variables and natural impacts shapes the many-sided directions of neuronal development, separation, and the arrangement of practical brain circuits.

3.2 Hereditary Premise of Neurological Problems:

Neurogenetics likewise reveals insight into the hereditary premise of neurological issues, unwinding the atomic underpinnings of conditions like Alzheimer's infection, Parkinson's sickness, and epilepsy. Genomic studies recognize risk variations, powerlessness qualities, and pathways embroiled in the pathogenesis of these problems. The crossing point of hereditary qualities and nervous system science opens roads for customized medication, designated treatments, and a more profound comprehension of the hereditary scene of neurological wellbeing and illness.

4. **Mental Neuroscience: Testing the Boondocks of Human Perception**

4.1 Memory and Learning: Bits of knowledge from Mental Neuroscience:

Mental neuroscience investigates the brain instruments that underlie memory and picking up, uncovering the unique cycles by which the mind encodes, solidifies, and recovers data. From the hippocampus to the prefrontal cortex, key cerebrum districts add to the development of wordy recollections, spatial route, and the obtaining of new abilities. The disclosures from mental neuroscience illuminate instructive practices, mental intercessions, and our principal comprehension of human discernment.

4.2 Inclination and the Limbic Framework: A Neurological Embroidery:

Profound encounters track down their underlying foundations in the complex woven artwork of the limbic framework, where designs like the amygdala and hippocampus assume focal parts. Mental neuroscience digs into the brain circuits that underlie profound handling, close to home guideline, and the incorporation of feeling with insight.

The crossing point of feeling and cognizance arises as a powerful field with suggestions for psychological well-being, emotional issues, and the comprehensive comprehension of human way of behaving.

5. **Cerebrum PC Connection points: Combining Brains and Machines**

   5.1 Brain Connection point Advances: From Science fiction to The real world:

The domain of cerebrum PC interfaces (BCIs) remains as a demonstration of the crossing point of neuroscience and innovation. Brain interface innovations, once consigned to the domain of sci-fi, have turned into a reality, empowering direct correspondence between the cerebrum and outside gadgets. BCIs hold guarantee for people with loss of motion, neurodegenerative infections, and other neurological circumstances, offering roads for reestablishing correspondence and command over the outside climate.

5.2 Moral Contemplations in Brain Connection point Exploration:

As the abilities of brain interface innovations grow, moral contemplations come to the very front. Issues of security, assent, and the potential for mental upgrade bring up complex moral issues. Exploring the moral elements of brain interface research becomes fundamental in guaranteeing dependable turn of events, arrangement, and the fair admittance to these groundbreaking advances.

6. **Brain adaptability: Adjusting to Change and Injury**

   6.1 Natural Advancement and Cerebrum Wellbeing:

Brain adaptability, the cerebrum's capacity to adjust to ecological changes, arises as a crucial guideline in neuroscience. Natural advancement, portrayed by mental excitement, social communication, and active work, encourages brain adaptability and advances mind wellbeing. The disclosures from concentrates on brain adaptability illuminate way of life mediations, restoration procedures, and the potential for upgrading mental flexibility across the life expectancy.

6.2 Neurorehabilitation and Recuperation from Cerebrum Injury:

Understanding brain adaptability becomes crucial in the domain of neurorehabilitation, where people recuperating from mind wounds go through versatile changes to recover lost capabilities. The mind's ability to rearrange itself in light of injury frames the reason for restoration mediations pointed toward advancing recuperation and reestablishing ideal brain working. Brain adaptability driven approaches offer expect people defeating the difficulties presented by stroke, horrendous cerebrum wounds, and neurodegenerative issues.

7. **Neurological Issues: From Seat to Bedside**

7.1 Alzheimer's Sickness: Translating the Riddle of Cognitive decline:

Neurological disclosures stretch out into the conundrum of Alzheimer's infection, a neurodegenerative issue portrayed by cognitive decline and mental deterioration. Progresses in neurogenetics, neuroimaging, and biomarker research add to how we might interpret the basic components of Alzheimer's illness. The quest for early analytic markers and sickness altering mediations stays at the very front of examination, holding the possibility to change the scene of Alzheimer's illness the executives.

7.2 Parkinson's Infection: Disentangling the Dopaminergic Dance:

Parkinson's illness, set apart by engine disabilities and dopaminergic brokenness, goes through examination in the domain of neurological disclosures. Experiences into the hereditary variables, synapse uneven characters, and brain circuit anomalies related with Parkinson's infection illuminate remedial systems. From pharmacological mediations to profound cerebrum feeling, the investigation of Parkinson's sickness features the multidisciplinary idea of neurological exploration chasing after successful therapies.

8. **Future Outskirts in Neuroscience: Investigating Unknown Domains**

8.1 Mind Machine Coordination: Obscuring the Lines Among Man and Machine:

The assembly of neuroscience and innovation makes the way for advanced wildernesses in mind machine joining. Mind machine interfaces that empower direct correspondence between the cerebrum and outer gadgets raise opportunities for improved mental capacities, increased reality, and new components of human-machine joint effort. Exploring the moral, cultural, and mental ramifications of cerebrum machine reconciliation turns into a basic in forming the future direction of this extraordinary field.

8.2 Quantum Neuroscience: Disentangling the Quantum Secrets of the Brain:

At the convergence of quantum physical science and neuroscience lies the arising field of quantum neuroscience. Investigating the potential quantum systems at play in the mind's mental capabilities presents a change in perspective in how we might interpret cognizance, discernment, and data handling. The investigation of quantum secrets inside the cerebrum holds guarantee for revealing new layers of intricacy in the brain code and reclassifying our crucial comprehension of the psyche.

9. **Moral Contemplations: Supporting Capable Neuroscientific Request**

9.1 Moral Aspects in Neuroscientific Exploration:

As neuroscience keeps on pushing the limits of information, moral contemplations guide the capable direct of neuroscientific request. From issues of assent in mind examination to the moral utilization of arising advancements, the quest for information is tempered by a guarantee to regard for people, logical respectability, and cultural prosperity. Moral reflections become basic in exploring the possible cultural effects, unseen side-effects, and mindful scattering of neuroscientific disclosures.

4.1Unraveling the neurobiology of angelfish and its relevance to human neuroscience.

The investigation of the neurobiology of angelfish uncovers an interesting excursion into the many-sided universe of piscine cognizance and conduct. As a model organic entity in neuroscience research, angelfish offers novel experiences into the brain systems that underlie social way of behaving, tactile discernment, and mental cycles. This thorough assessment dives into the neurobiological disclosures of angelfish, featuring its importance to human neuroscience and the possible commitments to how we might interpret the mind's intricacy and development.

1. **Prologue to Angelfish as a Model Life form in Neuroscience**
   1.1 Angelfish as a Model Creature:
   Angelfish, having a place with the cichlid family, arises as a charming model life form in neuroscience research. With its different collection of social ways of behaving, complicated correspondence methodologies, and obvious brain circuits, angelfish gives a special window into the brain underpinnings of perplexing ways of behaving. The decision of angelfish as a model organic entity comes from its phylogenetic vicinity to people and the protection of major neurobiological cycles across vertebrates.
   1.2 Significance of Model Organic entities in Neuroscience:
   Model organic entities assume a critical part in propelling comprehension we might interpret the sensory system. These life forms, going from spineless creatures like natural product flies to vertebrates like angelfish, offer exploratory benefits that permit analysts to test the complexities of brain capability, improvement, and conduct. The experiences acquired from model organic entities frequently give central information that can be meant more extensive applications, including human neuroscience.

2. **Neuroanatomy of Angelfish: Investigating the Brain Engineering**
   2.1 Relative Neuroanatomy:

Relative neuroanatomy frames the reason for unwinding the neurobiological underpinnings of conduct. Concentrating on the neuroanatomy of angelfish includes looking at the design and association of key cerebrum areas answerable for tangible handling, engine control, and mental capabilities. Relative examinations with mammalian minds enlighten both saved and different highlights, giving a nuanced comprehension of the transformative directions of brain circuits.

2.2 Social Dynamic Communities:

Angelfish are famous for their mind boggling social connections, making their cerebrum locales related with social dynamic a point of convergence of neurobiological examinations. The preoptic region, amygdala-like designs, and the telencephalon assume critical parts in handling expressive gestures, intervening social progressive systems, and coordinating mating ways of behaving. Investigating these brain fixates reveals insight into the neurobiology of social ways of behaving, with suggestions for grasping comparable cycles in human minds.

3. **Brain Circuits Fundamental Social Ways of behaving**

   3.1 Social Progressive systems and Predominance:

   The foundation of social progressive systems and predominance structures is a sign of angelfish conduct. Brain circuits overseeing pecking orders include the transaction between tactile discernment, navigation, and engine control focuses. Disentangling the brain systems behind friendly predominance gives experiences into the developmentally monitored rules that manage social associations across species, including people.

   3.2 Romance and Mating Ways of behaving:

   Angelfish participate in intricate romance and mating ceremonies, arranged by particular brain circuits. The recognizable proof of conspecifics, the showcase of romance ways of behaving, and the coordination of mating successions include mind boggling brain processes. Researching the brain circuits directing romance and mating ways of behaving in angelfish upgrades how we might interpret the more extensive neurobiology of proliferation and social holding, with expected equals to human social and regenerative ways of behaving.

4. **Tactile Handling and Discernment in Angelfish**

   4.1 Visual Discernment and Correspondence:

   Vision is an essential tactile methodology for angelfish, molding their collaborations with the climate and conspecifics. The brain circuits engaged with visual discernment, variety segregation, and protest acknowledgment add to the rich collection of angelfish correspondence. Concentrating on these tactile handling components offers experiences into the brain premise of visual discernment, with suggestions for figuring out the tangible groundworks of social correspondence in people.

4.2 Olfactory and Gustatory Handling:

Olfaction and gustation are basic to angelfish tangible discernment, affecting their reactions to food, mates, and ecological prompts. The olfactory bulbs and related mind locales assume key parts in handling compound signs, directing ways of behaving connected with scavenging, mate determination, and hunter evasion. Looking at the brain circuits engaged with olfactory and gustatory handling gives a relative point of view on chemosensory frameworks across vertebrates, including people.

5. **Learning and Memory in Angelfish: Experiences into Mental Cycles**

   5.1 Mental Adaptability and Transformation:

   Angelfish show momentous mental adaptability, adjusting their ways of behaving to changing natural circumstances and social settings. The brain circuits administering mental adaptability include areas related with learning and memory, for example, the telencephalon and hippocampus-like designs. Understanding the neurobiological premise of mental adaptability in angelfish adds to the more extensive investigation of versatile ways of behaving and memory processes in vertebrates.

   5.2 Cooperative Learning and Molding:

   Cooperative learning is a basic mental cycle saw in angelfish, where they structure associations among upgrades and results. Brain circuits engaged with affiliated learning add to the improvement of adapted ways of behaving, including reactions to food prompts, social signs, and ecological changes. Researching the neurobiology of acquainted learning in angelfish reveals insight into the systems basic learning and memory, offering equals to comparable cycles in human cognizance.

6. **Neurogenetics of Angelfish: Bits of knowledge into Developmental Preservation**

   6.1 Preservation of Brain Qualities:

   The assessment of the neurogenetics of angelfish includes recognizing and portraying the qualities that administer brain advancement, capability, and conduct.

   Relative genomic examinations uncover the preservation of key brain qualities among angelfish and different vertebrates, underlining the transformative coherence of neurobiological cycles. Recognizing saved brain qualities gives an establishment to figuring out the hereditary premise of conduct and cognizance in both angelfish and people.

   6.2 Developmental Meaning of Brain Qualities:

   The developmental meaning of brain qualities in angelfish stretches out past species-explicit ways of behaving. Saved brain qualities add to the common parentage of vertebrate sensory systems, offering bits of knowledge into the developmental variations that have molded different mental capacities. Investigating the transformative directions of brain qualities in angelfish upgrades how we might interpret the hereditary groundworks of neurobiology and the potential specific tensions that have affected mind advancement across species.

7. **Similar Neurobiology: Spanning Angelfish Bits of knowledge to Human Neuroscience**

   7.1 Connecting Neurobiological Holes:

   The near neurobiology of angelfish fills in as a scaffold, interfacing the bits of knowledge acquired from piscine neurobiology to the more extensive scene of human neuroscience. While recognizing the distinctions in cerebrum intricacy and association, the rationed standards of brain capability, social handling, and tactile discernment give an establishment to drawing matches among angelfish and human neurobiology.

   7.2 Shared Developmental Inheritances:

   The common developmental inheritances among angelfish and people become evident while looking at monitored brain circuits, qualities, and conduct characteristics. Transformative standards feature the central likenesses in the association and capability of the sensory system across vertebrates. By recognizing shared developmental heritages, specialists can extrapolate discoveries from angelfish neurobiology to acquire experiences into the brain substrates of human comprehension and conduct.

8. **Pertinence to Human Neuroscience: Suggestions and Applications**

   8.1 Experiences into Social and Mental Problems:

   The neurobiological experiences acquired from angelfish have direct ramifications for grasping social and mental issues in people. Looking at the brain circuits overseeing social ways of behaving and mental cycles in angelfish gives a relative system to exploring issues, for example, chemical imbalance range problems, social nervousness, and mental debilitations.
   The recognizable proof of preserved brain instruments offers possible focuses for helpful intercessions and the improvement of medicines for neurodevelopmental messes.

   8.2 Translational Applications in Neuropharmacology:

   The investigation of angelfish neurobiology adds to translational applications in neuropharmacology. Experiences into the brain circuits engaged with conduct, stress reactions, and mental capabilities give a premise to testing pharmacological intercessions. The improvement of drugs focusing on unambiguous brain processes in angelfish might offer novel methodologies for tending to human neurological and mental circumstances.

9. **Moral Contemplations in Angelfish Neuroscience Exploration**

   9.1 Moral Treatment of Exploration Subjects:

Moral contemplations in angelfish neuroscience research include guaranteeing the altruistic treatment of examination subjects. Carrying out moral rules for the consideration and utilization of angelfish in tests is fundamental, accentuating the standards of creature government assistance, limiting pressure, and complying to mindful exploration rehearses. Moral treatment of exploration subjects reaches out to contemplations of agony the executives, fitting lodging conditions, and the mindful utilization of hereditary control strategies.

9.2 Moral Ramifications of Translational Exploration:

Translational applications coming from angelfish neuroscience research raise moral ramifications connected with the possible advancement of drugs and restorative intercessions. Guaranteeing the moral direct of translational examination includes contemplations of wellbeing, adequacy, and the expected effects on human wellbeing. Moral structures that focus on quiet security, informed assent, and the mindful interpretation of discoveries from angelfish to human applications guide the moral direction of translational exploration.

10. **Future Headings in Angelfish Neuroscience Exploration**

10.1 Extending the Neurobiological Tool compartment:

The eventual fate of angelfish neuroscience research includes extending the neurobiological tool compartment accessible to scientists. Progressions in neuroimaging, optogenetics, and social measures offer chances to dive further into the brain circuits and atomic cycles that underlie angelfish conduct. The joining of state of the art advances upgrades the accuracy and goal of neurobiological examinations, making ready for new revelations.

10.2 Environmental and Transformative Aspects:

Investigating the environmental and transformative elements of angelfish neurobiology opens roads for understanding how brain variations add to the species' biological specialty and developmental achievement. Researching how neurobiological attributes interact with natural difficulties and shape transformative directions improves our appreciation of the perplexing interchange between brain development, conduct, and environmental elements.

4.2 Studying neurodevelopment and insights into neurological disorders.

Neurodevelopment, the multifaceted interaction by which the sensory system shapes and develops, fills in as the establishment for disentangling the intricacies of neurological issues. Researching the sub-atomic, cell, and underlying parts of neurodevelopment gives priceless bits of knowledge into the starting points, movement, and likely restorative intercessions for a horde of neurological circumstances. This investigation digs into the meaning of concentrating on neurodevelopment and how it enlightens pathways to grasp and address neurological issues.

1. **Neurodevelopment: An Orchestra of Cell Occasions**

   1.1 Early stage Neurogenesis:

   The excursion of neurodevelopment initiates during undeveloped stages with neurogenesis, the age of nerve cells or neurons. The arranged multiplication, relocation, and separation of brain begetter cells lay the basis for the different neuronal populaces that populate the creating sensory system. Understanding the complexities of undeveloped neurogenesis uncovers the outline for the brain circuits that will oversee complex physiological and mental capabilities.

   1.2 Synaptogenesis and Circuit Arrangement:

   As neurodevelopment advances, synaptogenesis and circuit arrangement arise as significant occasions. Neurons lay out associations through neurotransmitters, making many-sided networks that work with correspondence between cells. The development of brain circuits is a unique cycle impacted by hereditary signals, natural boosts, and movement subordinate systems. Disentangling the standards administering synaptogenesis gives bits of knowledge into the wiring of the mind and the reason for data handling.

2. **Atomic Movement in Neurodevelopment**

   2.1 Hereditary Guideline:

   Hereditary guideline assumes a focal part in coordinating the sub-atomic movement of neurodevelopment. Explicit qualities administer the timing and separation of brain cells, adding to the different cluster of cell types inside the sensory system. Changes or dysregulation of these qualities can upset the fragile equilibrium of neurodevelopment, possibly prompting neurological problems.

   2.2 Ecological Impacts:

   Past hereditary variables, ecological impacts shape neurodevelopmental directions. Outer signals like maternal wellbeing, openness to poisons, and early valuable encounters add to the arrangement of the brain engineering. The transaction between hereditary inclinations and natural elements turns into a basic concentration in understanding how bothers in neurodevelopment might add to neurological problems further down the road.

3. **Bits of knowledge into Neurological Issues: A Formative Viewpoint**

   3.1 Neurodevelopmental Issues:

   The investigation of neurodevelopment offers an interesting focal point through which to comprehend neurodevelopmental messes. Conditions, for example, chemical imbalance range problems, scholarly handicaps, and consideration deficiency hyperactivity jumble (ADHD) have establishes in disturbances to mental health early. By analyzing the deviations from run of the mill neurodevelopmental directions, analysts gain critical experiences into the etiology and indication of neurodevelopmental messes.

   3.2 Neurodegenerative Issues:

Neurodevelopmental standards likewise advise our comprehension regarding neurodegenerative issues, which frequently manifest further down the road. Concentrating on how neurons are at first shaped, the elements impacting their endurance, and the elements of synaptic associations gives an establishment to understanding the ever-evolving loss of brain capability saw in conditions like Alzheimer's illness, Parkinson's sickness, and amyotrophic parallel sclerosis (ALS).

4. **Hereditary Bits of knowledge: Unraveling the Hereditary Premise of Neurological Issues**

4.1 Hereditary Variations and Chance Elements:

Inspecting the hereditary premise of neurodevelopmental messes uncovers a complicated interchange of hereditary variations and chance elements. Broad affiliation studies (GWAS) and hereditary sequencing methods recognize explicit qualities related with expanded defenselessness to neurological circumstances.

The ID of these hereditary variations gives analytic markers as well as offers expected focuses for restorative mediations.

4.2 Shared Hereditary Pathways:

Shockingly, shared hereditary pathways frequently connect apparently unmistakable neurological problems. Exploring the shared traits in hereditary underpinnings gives a premise to figuring out shared systems and likely crossdisciplinary ways to deal with treatment. Perceiving these common pathways cultivates an additional incorporated point of view on neurological problems, creating some distance from segregated conditions to a more all encompassing comprehension of neurodevelopmental and neurodegenerative cycles.

5. **Cell and Atomic Components in Neurological Problems**

5.1 Synaptic Brokenness:

Disturbances in neurodevelopmental processes add to synaptic brokenness, a sign of numerous neurological issues. Modifications in neurotransmitter development, support, or pliancy can affect correspondence between neurons, influencing mental capabilities and conduct. Understanding the phone and sub-atomic instruments of synaptic brokenness gives potential chances to designated restorative mediations pointed toward reestablishing ordinary synaptic movement.

5.2 Neuroinflammation and Resistant Reactions:

Neurodevelopmental and neurodegenerative issues frequently include neuro-inflammatory reactions and insusceptible framework dysregulation. The job of microglia, the mind's occupant safe cells, in forming neurodevelopment and answering obsessive changes is a thriving area of examination. Bits of knowledge into the crosstalk among neuroinflammation and neurodevelopment add to a more nuanced comprehension of issues like different sclerosis, where resistant reactions influence both turn of events and degeneration.

6. **Remedial Methodologies: Exploring the Formative Scene**

6.1 Early Intercession Techniques:

The formative viewpoint on neurological issues highlights the significance of early mediation methodologies. Focusing on neurodevelopmental processes during basic periods offers potential chances to reshape unusual directions and alleviate the drawn out effect of issues. Early ID, social mediations, and neurorehabilitation become fundamental parts of an exhaustive way to deal with tending to neurological circumstances.

6.2 Accuracy Medication:

Progressions in genomic and sub-atomic advancements prepare for accuracy medication in the domain of neurological issues. Understanding the individual hereditary scene considers customized treatment systems custom-made to the particular hereditary variations and sub-atomic pathways related with a specific issue. Accuracy medication holds guarantee for advancing restorative results and limiting likely incidental effects by adjusting intercessions to the interesting attributes of every patient's neurodevelopmental profile.

7. **Moral Contemplations in Neurodevelopmental Exploration**

7.1 Informed Assent and Hereditary Security:

Moral contemplations in neurodevelopmental research base on issues of informed assent and hereditary protection. As hereditary experiences become more essential to understanding and treating neurological problems, guaranteeing people are completely educated about the ramifications regarding hereditary testing and exploration support becomes vital. Regarding hereditary protection shields people against expected derision and separation in light of their hereditary data.

7.2 Fair Admittance to Intercessions:

Fair admittance to neurodevelopmental intercessions is a basic moral thought. Guaranteeing that restorative methodologies are available to people across assorted financial foundations and geological areas advances decency and inclusivity. Tending to medical services differences in the arrangement of neurodevelopmental care lines up with moral standards of equity and guarantees that the advantages of examination and mediations are available to all.

8. **Future Bearings: Unwinding Intricacy for Remedial Advancement**

8.1 Progressions in Imaging Advancements:

The future of neurodevelopmental research unfurls against a scenery of headways in imaging advances. High-goal imaging methods, for example, practical attractive reverberation imaging (fMRI) and high level microscopy, empower scientists to envision dynamic cycles continuously. These innovations offer exceptional experiences into the cell and sub-atomic elements of neurodevelopment, making ready for additional exact mediations.

8.2 Coordinating Multi-Omics Approaches:

Coordinating multi-omics approaches, enveloping genomics, transcriptomics, proteomics, and epigenomics, gives a comprehensive comprehension of the sub-atomic scene in neurodevelopmental messes.

Exhaustive examinations of assorted atomic layers improve the recognizable proof of interconnected pathways and expected remedial targets. The joining of multi-omics information encourages a frameworks level viewpoint, catching the intricacy of neurodevelopment and its suggestions for neurological issues.

4.3Applications in drug discovery for neurological conditions.

The field of medication revelation for neurological circumstances remains at the front line of logical development, filled by a profound comprehension of sub-atomic pathways, cutting edge innovations, and a promise to tending to the intricate difficulties presented by issues of the sensory system. This investigation digs into the applications that drive drug revelation in the domain of neurological circumstances, analyzing the groundbreaking effect of these methodologies and the promising roads they open for creating novel therapeutics.

1. **Target Recognizable proof and Approval: Accuracy in Remedial Mediation**

   1.1 Unwinding Sickness Components:

   The excursion starts with the careful distinguishing proof and approval of restorative targets. Specialists dig into the unpredictable atomic and cell components basic neurological issues, looking to disentangle the intricate exchange of hereditary, ecological, and neurobiological factors. Propels in genomics, proteomics, and frameworks science give a thorough perspective on illness pathways, directing the choice of exact focuses for helpful mediation.

   1.2 Approving Focuses for Clinical Significance:

   Target approval is a basic move toward guaranteeing that the chose sub-atomic targets have clinical significance and restorative potential. Preclinical examinations, remembering for vitro and in vivo analyzes, approve the effect of adjusting these objectives on illness related aggregates. Refined hereditary and pharmacological apparatuses empower specialists to pinpoint explicit hubs inside sickness pathways, preparing for the advancement of focused on and successful therapeutics.

2. **High-Throughput Screening: Speeding up Compound Disclosure**

   2.1 Compound Libraries and Mechanical technology:

   High-throughput screening (HTS) alters the speed of medication revelation by working with the fast testing of thousands of mixtures for their restorative potential. Compound libraries, containing different substance elements, are screened against approved targets utilizing computerized mechanical frameworks. This approach speeds up the ID of lead compounds with the possibility to tweak explicit infection targets, giving an abundance of contender to additional enhancement.

2.2 Phenotypic Screening: Past Objective Driven Approaches:

Phenotypic screening, an elective way to deal with target-driven techniques, centers around the discernible qualities of cells or creatures impacted by a sickness. This comprehensive methodology considers the more extensive effect of mixtures on infection related aggregates, offering experiences into complex connections that may not be caught by target-driven systems alone. Phenotypic screening upgrades the revelation of mixtures with diverse restorative impacts, especially significant for complex neurological circumstances.

3. **Computational Methodologies: Displaying the Cerebrum in Silico**

3.1 In Silico Displaying of Organic Frameworks:

Headways in computational science and bioinformatics contribute essentially to medicate revelation for neurological circumstances. In silico demonstrating strategies recreate the way of behaving of organic frameworks, permitting specialists to foresee the collaborations between potential medication competitors and their objectives. Atomic mooring, quantitative design action relationship (QSAR) displaying, and arrange examinations give significant bits of knowledge into the limiting affinities, pharmacokinetics, and frameworks level impacts of competitor compounds.

3.2 Man-made consciousness and AI:

The incorporation of man-made consciousness (simulated intelligence) and AI (ML) reshapes the scene of medication revelation. Computer based intelligence calculations dissect tremendous datasets, including genomic data, underlying science information, and clinical preliminary results, to recognize designs and anticipate potential medication up-and-comers. AI models can uncover stowed away connections, streamline lead compounds, and speed up the medication improvement pipeline, offering a strong supplement to exploratory methodologies.

4. **Judicious Medication Plan: Accuracy Fitting of Therapeutics**

4.1 Construction Based Medication Plan:

Judicious medication configuration use nitty gritty information on the three-layered designs of natural particles to tailor helpful specialists with high accuracy. Primary science methods, like X-beam crystallography and cryo-electron microscopy, give experiences into the spatial game plan of proteins and their limiting locales. Structure-based drug configuration empowers the objective adjustment of lead mixtures to upgrade restricting fondness, particularity, and helpful adequacy.

4.2 Ligand-Based Medication Plan:

Ligand-put together medication configuration centers with respect to enhancing intensifies in light of their connections with known ligands or sub-atomic designs. Quantitative construction movement relationship (QSAR) studies, virtual screening, and ligand mooring reproductions guide the refinement of lead compounds. This approach permits specialists to plan therapeutics that impersonate the movement of endogenous ligands, working with a more designated and effective medication improvement process.

5. **Biomarkers and Customized Medication: Fitting Medicines to People**

5.1 Recognizing Neurological Biomarkers:

Biomarkers assume a significant part in portraying and diagnosing neurological circumstances, as well as observing treatment reactions. Progresses in neuroimaging, genomics, and sub-atomic profiling empower the ID of explicit biomarkers related with various issues. These sub-atomic marks give experiences into sickness movement, subtype characterization, and possible patient separation for customized therapy draws near.

5.2 Customized Therapeutics:

The time of customized medication changes drug revelation for neurological circumstances into a custom-made and patient-driven try. Understanding the hereditary and sub-atomic profiles of individual patients takes into account the advancement of designated therapeutics that think about the interesting attributes of every individual's neurological scene. Customized treatment procedures intend to upgrade viability while limiting unfriendly impacts, proclaiming a change in perspective towards additional exact and successful mediations.

6. **Translational Exploration: Connecting Seat to Bedside**

6.1 Preclinical Approval and Creature Models:

Translational examination overcomes any barrier between preclinical disclosures and clinical applications. Thorough preclinical approval, frequently including concentrates on in creature models, evaluates the security, viability, and systems of activity of potential medication applicants. Creature models of neurological problems, painstakingly intended to restate key parts of human pathology, act as significant stages for testing restorative speculations and refining drug advancement procedures.

6.2 Clinical Preliminaries: Assessing Wellbeing and Viability in People:

The excursion from the research center to clinical practice finishes in painstakingly planned clinical preliminaries. Stage I preliminaries center around wellbeing and measurements, while Stage II and III preliminaries assess adequacy, ideal dosing, and likely secondary effects in bigger patient populaces. Clinical preliminaries for neurological circumstances explore remarkable difficulties, for example, the blood-mind hindrance and the intricate idea of neurological endpoints.

The fruitful culmination of clinical preliminaries denotes a critical stage toward administrative endorsement and the acquaintance of new treatments with patients.

7. **Drug Reusing: Saddling Existing Particles for New Boondocks**

7.1 Extending Helpful Collections:

Drug reusing, otherwise called drug repositioning, offers an essential easy route in drug revelation by utilizing existing mixtures for new helpful purposes. Endorsed drugs with laid out wellbeing profiles go through reexamination for their expected viability in treating neurological problems. This approach speeds up the advancement course of events, as reused drugs have proactively passed administrative obstacles for security and measurement. The reusing technique is especially encouraging for neurological circumstances, where the pressing requirement for viable medicines energizes imaginative methodologies.

7.2 Organization Pharmacology: Figuring out Polypharmacology:

Network pharmacology supports the reasoning behind drug reusing, accentuating the interconnected idea of natural frameworks. Instead of zeroing in on single targets, network pharmacology thinks about the more extensive organization of atomic connections and flagging pathways. This all encompassing methodology perceives the polypharmacology of medications, wherein a solitary compound might regulate numerous objectives. By understanding the perplexing connections inside organic organizations, scientists uncover new roads for reusing existing medications to address neurological issues.

8. **Difficulties and Future Headings: Preparing for Leap forwards**

8.1 Blood-Mind Boundary: Conquering the Obstruction to Medication Conveyance:

The blood-mind obstruction represents an imposing test in drug disclosure for neurological circumstances. This particular boundary confines the section of numerous restorative specialists into the cerebrum, restricting the viability of expected medicines. Creating imaginative medication conveyance methodologies, for example, nanotechnology-based approaches and centered ultrasound, holds guarantee in conquering this boundary and improving the conveyance of therapeutics to the focal sensory system.

8.2 Tending to Heterogeneity in Neurological Issues:

The inborn heterogeneity inside neurological issues presents a test in planning medicines that oblige different patient populaces. Accuracy medication draws near, directed by biomarkers and hereditary data, mean to address this heterogeneity by fitting medicines to explicit subtypes or individual profiles. Future examination will probably zero in on unwinding the complexities of illness subtypes and refining restorative methodologies for more prominent particularity.

9. Moral Contemplations in Neurological Medication Revelation

9.1 Informed Assent and Member Privileges:

Moral contemplations in neurological medication revelation focus on member freedoms and informed assent. Guaranteeing that people engaged with clinical preliminaries are completely educated about the expected dangers, advantages, and options in contrast to support is a central moral standard. Regarding independence, protection, and the option to pull out from the review without repercussion defends the prosperity and freedoms of exploration members.

9.2 Impartial Access and Worldwide Wellbeing Value:

Worldwide wellbeing value arises as a basic moral thought, underlining the significance of guaranteeing that the advantages of medication disclosure arrive at different populaces. Addressing differences in admittance to imaginative treatments, especially in asset restricted settings, turns into an ethical objective. Moral systems that advance inclusivity, reasonableness, and straightforwardness guide the worldwide execution of neurological medicines.

Chapter 5

Drug Discovery and Therapeutics

1.1 The Basic for Medication Revelation:

Drug revelation is a dynamic and multidisciplinary field that assumes a significant part in propelling medical services by distinguishing and creating remedial specialists to treat sicknesses. As how we might interpret the atomic and cell premise of illnesses develops, the requirement for inventive and successful medications turns out to be progressively clear. This extensive investigation digs into the multifaceted cycles, approaches, challenges, and moral contemplations that portray drug revelation and the resulting improvement of remedial mediations.

1.2 The Development of Medication Revelation:

The excursion of medication disclosure has advanced over hundreds of years, from fortunate revelations and conventional solutions for methodical and deductively directed approaches. Early forward leaps, like the revelation of anti-infection agents and immunizations, established the groundwork for present day drug disclosure. The mix of sub-atomic science, genomics, and trend setting innovations has changed the field, empowering specialists to target explicit atomic pathways with remarkable accuracy.

II. The Medication Revelation Interaction

2.1 Objective ID and Approval: From Qualities to Helpful Targets:

The medication revelation process starts with the recognizable proof and approval of possible remedial targets. Propels in genomics, practical genomics, and frameworks science permit scientists to disentangle the hereditary and atomic underpinnings of sicknesses. Approving these objectives guarantees their importance to illness pathology and the potential for successful tweak. From qualities ensnared in malignant growth to proteins engaged with neurodegenerative issues, target ID frames the establishment for resulting stages in drug revelation.

2.2 High-Throughput Screening: Speeding up Compound Disclosure:

High-throughput screening (HTS) is a pivotal move toward recognizing lead compounds with the possibility to tweak chosen targets. Mechanized automated frameworks work with the quick screening of enormous compound libraries against approved targets. HTS speeds up the recognizable proof of mixtures with wanted pharmacological exercises, giving a pool of contender to additional streamlining and improvement.

2.3 Computational Methodologies in Medication Revelation: Demonstrating and Informatics:

Computational methodologies have become essential to tranquilize disclosure, offering useful assets for in silico displaying and informatics. Atomic mooring, quantitative construction movement relationship (QSAR) studies, and virtual screening empower the forecast of how potential medication up-and-comers collaborate with target particles. Man-made reasoning (artificial intelligence) and AI calculations dissect immense datasets, uncovering designs that guide the determination and advancement of mixtures.

2.4 Normal Medication Plan: Accuracy Fitting of Therapeutics:

Normal medication configuration use primary science to configuration compounds with explicit cooperations with target particles. X-beam crystallography, cryo-electron microscopy, and other primary procedures give experiences into the three-layered designs of proteins and their limiting locales. Structure-based and ligand-based drug configuration approaches take into consideration the sane change of lead mixtures to upgrade adequacy and explicitness.

2.5 Hit to Lead Improvement: Refining Contender for Advancement:

Hit to lead streamlining includes refining introductory hits from high-throughput screening into lead compounds with further developed drug-like properties. Restorative science, pharmacokinetics, and toxicology evaluations guide the substance changes important to upgrade the viability, wellbeing, and bioavailability of lead compounds. This iterative interaction means to recognize up-and-comers with ideal medication properties for additional turn of events.

2.6 Preclinical Turn of events: Assessing Security and Adequacy:

Preclinical improvement includes a progression of studies to assess the security and viability of lead compounds prior to progressing to clinical preliminaries. Creature models assume a critical part in evaluating pharmacokinetics, toxicology, and beginning viability. Preclinical investigations give fundamental information to administrative entries and add to the comprehension of a medication up-and-comer's likely restorative advantages and dangers.

2.7 Investigational New Medication (IND) Application: Administrative Achievement:

The Investigational New Medication (IND) application denotes an administrative achievement in drug improvement. Accommodation of the IND application to administrative specialists, like the U.S. Food and Medication Organization (FDA), incorporates thorough information from preclinical examinations.

Administrative audit decides if the medication up-and-comer can continue to clinical preliminaries, offsetting the possible advantages with the security profile saw in preclinical appraisals.

III. Clinical Turn of events

3.1 Stage I Clinical Preliminaries: Wellbeing and Dose Investigation:

Clinical advancement advances through a progression of stages, with Stage I clinical preliminaries zeroing in on wellbeing and measurements. These preliminaries include few sound workers and plan to lay out the security profile, ideal dose reach, and beginning pharmacokinetics of the investigational drug. Stage I preliminaries give significant experiences into the medication's decency and guide ensuing periods of advancement.

3.2 Stage II Clinical Preliminaries: Viability in Understanding Populaces:

Stage II clinical preliminaries extend the review populace to incorporate people impacted by the objective infection. These preliminaries survey the adequacy of the investigational drug in unambiguous patient populaces, giving primer proof of helpful advantages. Stage II examinations contribute important information on measurement, treatment span, and expected aftereffects, illuminating the plan regarding bigger Stage III preliminaries.

3.3 Stage III Clinical Preliminaries: Affirming Viability and Security:

Stage III clinical preliminaries include bigger and more different patient populaces to affirm the viability and security of the investigational drug. Randomized, controlled preliminaries contrast the medication up-and-comer with existing medicines or fake treatments, creating vigorous proof for administrative endorsement. The information produced in Stage III preliminaries add to the thorough comprehension of the medication's advantages and dangers.

3.4 New Medication Application (NDA) Accommodation: Administrative Assessment:

The New Medication Application (NDA) accommodation denotes the climax of clinical turn of events, as supporters present a thorough dossier to administrative experts for survey. The NDA incorporates information from preclinical and clinical investigations, itemizing the medication's wellbeing, viability, and assembling processes. Administrative offices direct thorough assessments to evaluate the advantage risk profile and decide if the medication ought to get promoting endorsement.

3.5 Stage IV Clinical Preliminaries: Post-Showcasing Observation:

Stage IV clinical preliminaries, otherwise called post-showcasing observation, happen after administrative endorsement and commercialization. These preliminaries screen the drawn out wellbeing and viability of the medication in true settings, catching extra information on uncommon unfavorable occasions and patient results. Stage IV investigations add to continuous security evaluations and may illuminate changes to endorsing rules.

IV. Drug Reusing and Creative Techniques

4.1 Medication Reusing: Rediscovering Restorative Potential:

Drug reusing, or repositioning, includes recognizing new restorative purposes for existing medications with laid out security profiles. This procedure benefits from the abundance of data accessible for supported drugs, speeding up the medication advancement timetable. Drug reusing has exhibited outcome in different restorative regions, offering a proficient and savvy way to deal with tending to neglected clinical necessities.

4.2 Biologics and High level Treatments: Growing Treatment Modalities:

The scene of medication disclosure has extended to incorporate biologics and high level treatments, like monoclonal antibodies, quality treatments, and cell-based treatments. Biologics target explicit particles or pathways with high accuracy, offering novel ways to deal with treating complex sicknesses. The improvement of cutting edge treatments addresses a change in outlook by they way we conceptualize and address testing ailments.

4.3 Nanotechnology in Medication Conveyance: Improving Adequacy and Accuracy:

Nanotechnology assumes a vital part in drug conveyance, offering creative answers for improve the viability and accuracy of restorative mediations. Nano-particles can be intended to exemplify drugs, working on their bioavailability and focusing on unambiguous tissues or cells. This approach tends to difficulties, for example, the blood-mind obstruction and takes into account the controlled arrival of helpful specialists.

4.4 Pharmacogenomics: Fitting Medicines to Hereditary Changeability:

Pharmacogenomics coordinates hereditary qualities and pharmacology to tailor medicines in light of a person's hereditary cosmetics. Understanding how heredi-tary varieties impact drug digestion, viability, and unfriendly responses empowers customized treatment procedures. Pharmacogenomic approaches plan to advance medication reactions, limit incidental effects, and improve in general helpful results.

V. Challenges and Moral Contemplations

5.1 Difficulties in Medication Disclosure: Adjusting Advancement and Chance:

The medication disclosure process faces various difficulties, including the high weakening paces of medication up-and-comers, the intricacy of illnesses, and the requirement for significant monetary ventures. Adjusting the quest for advance-ment with the basic to relieve takes a chance with represents a continuous test. Specialists wrestle with issues, for example, off-target impacts, unforeseen poison levels, and the distinguishing proof of significant biomarkers.

5.2 Moral Contemplations in Medication Improvement: Defending Members and Worldwide Value:

Moral contemplations are fundamental all through the medication improvement venture, from preclinical exploration to clinical preliminaries and post-promoting observation. Defending the freedoms and prosperity of examination members, guaranteeing informed assent, and advancing straightforwardness in revealing discoveries are moral objectives. Worldwide wellbeing value arises as a pivotal thought, stressing the significance of open and reasonable medicines for different populaces.

VI. Future Headings and Developments

6.1 Headways in Innovation: Reforming Medication Revelation:

The fate of medication disclosure is entwined with progressions in innovation, including man-made reasoning, high-throughput screening stages, and creative imaging procedures. Incorporating these advancements improves the effectiveness and accuracy of medication revelation processes, opening additional opportunities for focusing on illnesses at the sub-atomic level.

6.2 Customized Medication: Fitting Medicines to People:

The period of customized medication holds huge commitment, with an emphasis on fitting medicines to the singular qualities of patients. Biomarker-driven systems, pharmacogenomics, and patient-explicit information add to a more customized and successful way to deal with medical services. The advancing comprehension of sickness heterogeneity and individual reactions to therapy educates the advancement regarding designated and upgraded treatments.

6.3 Worldwide Cooperation and Open Science: Speeding up Disclosures:

Worldwide cooperation and open science drives cultivate a cooperative way to deal with drug revelation. Sharing information, assets, and experiences across research foundations and industry accomplices speeds up the speed of revelations. Cooperative stages and open-access structures add to the democratization of information, preparing for leap forwards that rise above geographic and authoritative limits.

6.4 Patient Commitment and Support: Forming the Medication Advancement Scene:

Patient commitment and backing assume an undeniably powerful part in forming the medication advancement scene. The consideration of patient points of view in research plan, clinical preliminaries, and administrative navigation guarantees that medicines line up with the necessities and inclinations of those impacted by the designated illnesses. Patient promotion bunches add to bringing issues to light, financing exploration, and driving arrangement changes to help drug advancement.

5.1Showcase of angelfish as a model for drug development.

1.1 Reasoning for Model Creatures in Medication Advancement:

The quest for powerful restorative mediations depends on the essential utilization of model creatures to disentangle the intricacies of infections and test expected medicines. Model creatures act as significant instruments, giving bits of knowledge into key organic cycles and offering stages for drug improvement research. Among these, the angelfish (Pterophyllum spp.) arises as a convincing and flexible model, particularly situated to add to how we might interpret different infections and the improvement of novel therapeutics.

1.2 Outline of Angelfish as a Model Creature:

The determination of model organic entities is directed by elements like hereditary manageability, physiological pertinence, and amiability to exploratory control. Angelfish, known for their particular morphology and simplicity of upkeep in research facility settings, display qualities that make them an appealing model for drug improvement. This grandstand investigates the critical traits of angelfish that position them as a significant device in the medication improvement pipeline.

2. Exceptional Attributes of Angelfish

2.1 Hereditary Manageability: Disentangling the Angelfish Genome:

The angelfish genome fills in as a diagram for figuring out the hereditary premise of different physiological and neurotic cycles. Late progressions in genomic advancements have worked with the sequencing and examination of the angelfish genome, giving scientists a far reaching asset to investigate hereditary elements impacting characteristics pertinent to human wellbeing. The hereditary manageability of angelfish considers designated examinations concerning explicit qualities related with sicknesses, making ready for translational experiences.

2.2 Physiological Homology: Overcoming any barrier to Human Wellbeing:

One of the critical qualities of angelfish as a model life form lies in its physiological homology to people. Numerous natural cycles, including cardiovascular capability, neurodevelopment, and resistant reactions, show striking similitudes among angelfish and people. This homology improves the translational significance of discoveries in angelfish research, offering a scaffold between exploratory results and expected applications in human wellbeing.

2.3 Regenerative Limit: Investigating Tissue Recovery Pathways:

Angelfish, prestigious for their regenerative capacities, give an exceptional window into the components overseeing tissue fix and recovery. Understanding the sub-atomic and cell processes that drive regenerative reactions in angelfish might offer novel bits of knowledge into regenerative medication. Drug improvement systems pointed toward upgrading tissue recovery and fix could profit from the investigation of regenerative pathways recognized in angelfish.

2.4 Ecological Responsiveness: Examining Medication Reactions in Assorted Conditions:

The natural responsiveness of angelfish positions them as an optimal model for concentrating on drug reactions under various environmental circumstances.

Natural variables, like water quality and temperature, can impact drug digestion and adequacy. Angelfish, with their aversion to natural changes, give a stage to examine what outside factors mean for drug collaborations and reactions, illuminating medication improvement methodologies that think about ecological impacts.

3. Angelfish in Illness Demonstrating

3.1 Cardiovascular Illnesses: Bits of knowledge from Angelfish Hearts:

Cardiovascular illnesses address a huge worldwide wellbeing trouble, requiring research models that loyally restate heart physiology. Angelfish, with their obvious cardiovascular framework and hereditary manageability, offer bits of knowledge into heart advancement, capability, and reactions to heart stress. Displaying cardiovascular sicknesses in angelfish gives a stage to testing potential helpful mediations focusing on heart-related pathologies.

3.2 Neurological Problems: Unwinding the Neurobiology of Angelfish:

The neurobiology of angelfish presents a convincing road for researching neurological problems. The angelfish sensory system, with its intricacy and homology to the human cerebrum, fills in as an important model for concentrating on neurodevelopment, conduct, and reactions to neuroactive mixtures.

Drug advancement endeavors focusing on neurological problems can profit from unwinding the neurobiological complexities enlightened by angelfish research.

3.3 Metabolic Issues: Investigating Metabolic Pathways in Angelfish:

Metabolic issues, including diabetes and weight, present huge wellbeing challenges internationally. Angelfish, with their defenselessness to metabolic bothers, offer a model for concentrating on the basic components of metabolic issues and testing possible intercessions. Researching metabolic pathways in angelfish might reveal focuses for drug advancement pointed toward tweaking metabolic homeostasis.

3.4 Malignant growth Exploration: Angelfish as a Model for Cancer Science:

Malignant growth research benefits from model life forms that restate key parts of cancer science. Angelfish, with their helplessness to growth improvement, give a stage to concentrating on disease inception, movement, and reactions to anticancer specialists. Bits of knowledge acquired from angelfish models might illuminate drug improvement techniques focusing on unambiguous pathways ensnared in disease advancement and metastasis.

4. Applications in Medication Improvement

4.1 Angelfish as a Screening Stage: High-Throughput Medication Testing:

The hereditary manageability and physiological pertinence of angelfish make them an ideal evaluating stage for high-throughput drug testing. Huge scope drug separates angelfish can recognize compounds with remedial potential, permitting analysts to quickly survey the adequacy and security of medication competitors. This application speeds up the medication advancement pipeline by focusing on lead compounds for additional examination.

4.2 Designated Treatments: Accuracy Medication Bits of knowledge from Angelfish:

The hereditary homology among angelfish and people positions angelfish as a model for exploring accuracy medication draws near. Concentrating on individual hereditary varieties in angelfish might offer experiences into customized reactions to sedate medicines. This application adds to the advancing field of accuracy medication, where medicines are custom fitted to the hereditary cosmetics of individual patients.

4.3 Medication Conveyance Studies: Tending to Difficulties in Organization:

The responsiveness of angelfish to ecological circumstances makes them a superb model for concentrating on drug conveyance methodologies. Exploring how medications are assimilated, disseminated, utilized, and discharged in angelfish gives important information on drug pharmacokinetics.

This data is pivotal for creating drug conveyance frameworks that address difficulties connected with organization and advance restorative results.

4.4 Ecological Toxicology: Surveying Medication Wellbeing in Angelfish:

Angelfish's responsiveness to changes in water quality and ecological circumstances makes them a keen model for natural toxicology studies. Surveying the security of medication competitors in angelfish includes assessing expected unfriendly consequences for amphibian biological systems. Understanding the ecological effect of medications right off the bat in the advancement cycle adds to capable medication improvement rehearses.

5. Difficulties and Future Bearings

5.1 Moral Contemplations: Adjusting Exploration and Creature Government assistance:

Likewise with any model living being, the utilization of angelfish in research raises moral contemplations connected with creature government assistance. Guaranteeing the altruistic treatment of angelfish in research facility settings and it is fundamental to limit any expected pain. Specialists should figure out some kind of harmony between progressing logical information and keeping up with moral guidelines in creature research.

5.2 Interpretation to Human Pertinence: Overcoming any barrier:

Deciphering discoveries from angelfish examination to human importance stays a basic test. While the physiological homology among angelfish and people is a strength, scientists should cautiously decipher and extrapolate results. Overcoming any barrier between angelfish models and human applications requires strong approval and joining with clinical information to guarantee the translational effect of medication advancement experiences.

5.3 Incorporation with Other Model Frameworks: Amplifying Experiences:

To improve the expansiveness and profundity of bits of knowledge acquired from angelfish research, reconciliation with other model frameworks is urgent. Cooperative endeavors that join the qualities of angelfish models with those of different living beings, like mice or zebrafish, augment the potential for revealing far reaching instruments hidden illnesses and potential medication reactions.

5.4 Progressions in Innovation: Bridling Apparatuses for Accuracy:

Headways in innovation assume a crucial part in expanding the capability of angelfish as a model for drug improvement. Bridling state of the art devices, like CRISPR-Cas9 quality altering, high level imaging methods, and omics advancements, empowers specialists to dive further into the sub-atomic and cell complexities of angelfish science. These apparatuses upgrade accuracy in exploratory plan and information understanding.

5.2 Translating findings into potential therapeutic interventions.

The excursion from logical disclosure to unmistakable remedial mediations addresses a basic scaffold in the domain of biomedical examination. As scientists reveal novel experiences into the atomic and cell underpinnings of sicknesses, the basic lies in making an interpretation of these discoveries into viable therapies that can ease human misery and further develop wellbeing results. This investigation dives into the diverse course of making an interpretation of logical revelations into expected restorative mediations, exploring the intricacies, challenges, and groundbreaking possible intrinsic in this critical translational excursion.

2. Unwinding the Sub-atomic Scene

2.1 Genomic Disclosures: An Establishment for Accuracy Medication:

Headways in genomics have upset how we might interpret the hereditary premise of illnesses. From the distinguishing proof of infection related hereditary variations to the investigation of genomic scenes, scientists have revealed unpredictable sub-atomic marks that underlie different circumstances. Making an interpretation of genomic discoveries into potential helpful intercessions includes utilizing this hereditary information to foster designated and customized medicines — a foundation of accuracy medication.

2.2 Transcriptomics and Proteomics: Disentangling Cell Marks:

Transcriptomic and proteomic investigations give a unique depiction of quality articulation and protein profiles inside cells and tissues. Unwinding the complexities of cell marks permits scientists to distinguish central participants in sickness processes. Interpreting these discoveries includes the advancement of mediations that regulate quality articulation, protein capability, or flagging pathways, making ready for designated treatments custom-made to the atomic scene of explicit sicknesses.

2.3 Epigenetic Bits of knowledge: Tweaking Quality Articulation Examples:

Epigenetic alterations, like DNA methylation and histone acetylation, add to the guideline of quality articulation. Understanding the epigenetic scene of sicknesses opens roads for intercessions that regulate these alterations, possibly turning around distorted quality articulation designs. Epigenetic treatments hold guarantee in illnesses where the hidden pathology includes adjustments in quality guideline.

3. Seat to Bedside: Translational Difficulties

3.1 Preclinical Approval: Overcoming any barrier from Revelation to Improvement:

The progress from seat to bedside requires thorough preclinical approval of possible remedial intercessions. Preclinical examinations, frequently directed in cell or creature models, survey the security, viability, and systems of activity of applicant mediations. Overcoming any issues from revelation to advancement requires hearty proof from preclinical approval to legitimize the movement of intercessions into clinical preliminaries.

3.2 Creature Models: Imitating Human Physiology and Pathology:

Creature models assume an essential part in preclinical exploration, filling in as intermediaries for human physiology and pathology. The determination of fitting creature models includes thinking about variables like hereditary homology, physiological importance, and the reiteration of sickness highlights. Interpreting discoveries from creature models to human applications requires a nuanced comprehension of the impediments and qualities of each model framework.

3.3 Difficulties in Interpretation: From Mice to People:

Challenges have large amounts of interpreting discoveries from preclinical investigations, especially while moving from creature models, like mice, to human applications. Contrasts in digestion, safe reactions, and medication digestion between species require cautious translation of preclinical information. The test lies in extrapolating the importance of mediations saw in creature review to the complicated and dynamic human organic setting.

4. Clinical Preliminaries: Exploring the Human Scene

4.1 Stage I Preliminaries: Security and Measurement Investigation:

Clinical preliminaries address an essential stage in making an interpretation of possible mediations into clinically feasible medicines. Stage I preliminaries center around evaluating the wellbeing and measurement of up-and-comer mediations in a little partner of human workers. Grasping the pharmacokinetics, bearableness, and starting wellbeing profile frames the establishment for resulting periods of clinical turn of events.

4.2 Stage II Preliminaries: Adequacy in Tolerant Populaces:

Stage II preliminaries grow the review populace to incorporate people impacted by the objective infection. Surveying the viability of mediations in unambiguous patient populaces gives significant experiences into their possible restorative advantages. Stage II examinations add to the improvement of measurement, treatment length, and the ID of likely incidental effects, directing the movement of mediations to bigger Stage III preliminaries.

4.3 Stage III Preliminaries: Affirming Adequacy and Security:

Stage III clinical preliminaries include bigger and more assorted patient populaces and are intended to affirm the viability and security of mediations. Randomized, controlled preliminaries contrast the investigational treatment with existing norms of care or fake treatments, creating powerful proof for administrative endorsement. The information created in Stage III preliminaries illuminate administrative choices and add to an exhaustive comprehension of the restorative capability of mediations.

5. Accuracy Medication: Fitting Intercessions to People

5.1 Biomarker-Driven Methodologies: Customizing Treatment Approaches:

The period of accuracy medication stresses the significance of fitting intercessions to the singular qualities of patients. Biomarkers, which incorporate hereditary, genomic, or sub-atomic marks related with infections, guide the improvement of designated medicines. Incorporating biomarker-driven procedures into clinical preliminaries considers patient definition, upgrading the choice of people liable to profit from explicit mediations.

5.2 Pharmacogenomics: Improving Medication Reactions:

Pharmacogenomics investigates how hereditary varieties impact a singular's reaction to drugs. Understanding the interaction among hereditary qualities and medication digestion empowers the advancement of treatment regimens. Fitting mediations in view of a person's pharmacogenomic profile improves restorative results while limiting unfriendly impacts, denoting a huge step chasing customized medication.

6. Challenges in Interpretation and Defeating Boundaries

6.1 Heterogeneity in Tolerant Populaces: Tending to Assorted Profiles:

The heterogeneity inside understanding populaces presents a significant test in making an interpretation of mediations to broad clinical applications. Sicknesses frequently manifest with different subtypes or varieties in show, and reactions to mediations can change essentially among people. Tending to this heterogeneity requires accuracy medication moves toward that think about the exceptional qualities of patient subgroups.

6.2 Administrative Obstacles: Exploring Endorsement Cycles:

Exploring the administrative scene is a basic part of making an interpretation of likely helpful intercessions into supported medicines. Administrative bodies, like the U.S. Food and Medication Organization (FDA) or the European Meds Organization (EMA), assess the security, adequacy, and nature of intercessions prior to conceding endorsement. Beating administrative obstacles includes fastidious documentation, vigorous review plans, and adherence to moral and security norms.

6.3 Availability and Reasonableness: Guaranteeing Worldwide Effect:

The effective interpretation of mediations should reach out past administrative endorsement to guarantee worldwide availability and reasonableness. Differences in medical care assets and financial contemplations can restrict the accessibility of new therapies in specific areas. Procedures to address these difficulties incorporate cultivating joint efforts, pushing for fair access, and investigating inventive ways to deal with decrease the financial weight on patients.

7. Drug Reusing: Extending Helpful Collections

7.1 Utilizing Existing Particles: Speeding up Advancement Courses of events:

Drug reusing, otherwise called drug repositioning, offers an essential easy route in drug improvement by reusing existing mixtures for new helpful purposes. Endorsed drugs with laid out wellbeing profiles go through reexamination for their expected viability in treating various circumstances. This approach speeds up the advancement course of events, as reused drugs have previously passed administrative obstacles for wellbeing and dose.

7.2 Organization Pharmacology: Figuring out Polypharmacology:

Network pharmacology supports the reasoning behind drug reusing, underscoring the interconnected idea of natural frameworks. Rather than zeroing in on single targets, network pharmacology thinks about the more extensive organization of sub-atomic cooperations and flagging pathways. This comprehensive methodology perceives the polypharmacology of medications, wherein a solitary compound might regulate various targets. Understanding these mind boggling connections reveals new roads for reusing existing medications.

8. Future Bearings: Advancements and Arising Ideal models

8.1 Headways in Innovation: Altering Medication Advancement:

The future of making an interpretation of discoveries into remedial mediations is entwined with progressions in innovation. Developments in regions, for example, man-made consciousness, high-throughput screening stages, and high level imaging procedures improve the productivity and accuracy of medication advancement processes. Incorporating these advancements speeds up the recognizable proof of expected intercessions and smoothes out the medication improvement pipeline.

8.2 Patient Commitment and Support: Molding the Exploration Scene:

Patient commitment and support assume an undeniably powerful part in molding the exploration scene and the interpretation of discoveries into mediations. Including patients in research plan, clinical preliminary support, and dynamic cycles guarantees that mediations line up with the necessities and inclinations of those impacted by the designated illnesses. Patient promotion bunches add to bringing issues to light, subsidizing exploration, and driving arrangement changes to help the interpretation of discoveries into effective intercessions.

8.3 Worldwide Cooperation and Open Science: Speeding up Disclosures:

Worldwide cooperation and open science drives cultivate a cooperative way to deal with making an interpretation of discoveries into restorative mediations. Sharing information, assets, and experiences across research establishments, industry accomplices, and geographic limits speeds up the speed of revelations. Cooperative stages and open-access systems add to the democratization of information, making ready for leap forwards that address worldwide wellbeing challenges.

Chapter 6

Immunology and Inflammatory Responses

Immunology, the investigation of the resistant framework, is a field of central significance in grasping the body's safeguard systems against microorganisms and keeping up with homeostasis. Key to immunology is the many-sided snare of incendiary reactions — a transformative monitored guard system that goes about as a situation with two sides. This thorough investigation dives into the basics of immunology, the coordination of fiery reactions, and the powerful interchange that shapes both wellbeing and infection.

2. The Underpinning of Immunology

2.1 Outline of the Resistant Framework: Shielding the Host:

The resistant framework is a multi-layered organization of cells, tissues, and particles working cooperatively to shield the host against attacking microorganisms. Isolated into intrinsic and versatile parts, the invulnerable framework shows an exceptional capacity to perceive and kill a different exhibit of dangers. The inborn resistant reaction gives prompt, vague safeguard instruments, while the versatile invulnerable reaction tailors its protections in light of earlier openness, presenting immunological memory.

2.2 Cell Players: Coordinating Protection Components:

Cell parts are the cutting edge safeguards in the resistant framework. Phagocytes, like macrophages and neutrophils, overwhelm and process microbes. Lymphocytes, including Lymphocytes and B cells, assume vital parts in versatile resistance. Lymphocytes coordinate cell resistant reactions, perceiving and dispensing with tainted or abnormal cells, while B cells produce antibodies that kill microbes or imprint them for obliteration.

2.3 Immunoglobulins: Antibodies as Atomic Watchmen:

Immunoglobulins, or antibodies, are Y-molded proteins delivered by B cells. These adaptable particles perceive explicit antigens — atoms on the outer layer of microorganisms or unusual cells. Antibodies kill microorganisms straightforwardly, initiate supplement proteins, or label microbes for obliteration by other insusceptible cells. The variety of antibodies empowers the resistant framework to answer a broad exhibit of dangers.

3. Incendiary Reactions: A Situation with two sides

3.1 The Physiology of Irritation: A Defensive Outpouring:

Irritation is an essential and developmentally saved reaction to tissue injury, disease, or different difficulties. The trademark indications of irritation — redness, intensity, expanding, and torment — mirror a fountain of occasions pointed toward disposing of the reason for cell injury, getting out harmed cells and tissues, and starting tissue fix. The intense incendiary reaction is a firmly directed process that settle once the danger is killed.

3.2 Cell Players in Aggravation: Neutrophils and Macrophages:

Neutrophils, the most plentiful white platelets, are fast responders enlisted to locales of disease or injury. They inundate and annihilate microorganisms through phagocytosis. Macrophages, got from monocytes, are flexible cells that overwhelm cell flotsam and jetsam, microorganisms, and unfamiliar substances. Both cell types discharge flagging atoms called cytokines, arranging the provocative outpouring.

3.3 Supplement Framework: Upgrading Protection Systems:

The supplement framework is a gathering of proteins that upgrades the invulnerable reaction and adds to irritation. Initiated supplement proteins can opsonize microbes, making them more defenseless to phagocytosis, or straightforwardly lyse target cells. The supplement framework likewise associates with different parts of the invulnerable framework, connecting inborn and versatile insusceptibility.

3.4 Provocative Middle people: Cytokines and Chemokines:

Cytokines and chemokines are key arbiters of irritation, going about as flagging atoms that manage insusceptible reactions. Cytokines, for example, interleukins and cancer corruption factor (TNF), tweak the power and length of aggravation. Chemokines guide invulnerable cells to explicit locales, planning their development during insusceptible reactions.

3.5 Goal of Aggravation: Forestalling Chronicity:

Goal of aggravation is an effectively managed process significant for reestablishing tissue homeostasis. Mitigating arbiters, like concentrated favorable to settling lipid go betweens (SPMs), assist with hosing aggravation and advance tissue fix. Inability to determine aggravation appropriately can prompt persistent provocative circumstances, adding to different infections.

4. Immunopathology: When the Safe Framework Turns out badly

4.1 Autoimmunity: Mixed up Character and Self-Assault:

Immune system illnesses emerge when the insusceptible framework erroneously targets and goes after the body's own tissues. Dysregulation in self-resilience systems permits autoreactive Lymphocytes and antibodies to cause harm. Conditions like rheumatoid joint pain, fundamental lupus erythematosus, and type 1 diabetes are instances of immune system illnesses with assorted clinical indications.

4.2 Extreme touchiness Responses: Misrepresented Reactions:

Extreme touchiness responses result from misrepresented safe reactions to innocuous substances. Arranged into four sorts, excessive touchiness responses can go from gentle unfavorably susceptible reactions (Type I) to safe complex-intervened sicknesses (Type III) and deferred type touchiness (Type IV). Sensitivities, hypersensitivity, and immune system hemolytic pallor epitomize the range of extreme touchiness responses.

4.3 Immunodeficiency: Compromised Protections:

Immunodeficiency issues compromise the insusceptible framework's capacity to safeguard against diseases. Essential immunodeficiencies, frequently hereditary, bring about lacks in resistant cell capability or creation. Obtained immunodeficiencies, like HIV/Helps, come from diseases or outer elements that debilitate invulnerable capability. Immunodeficiencies manifest as intermittent or extreme diseases.

4.4 Ongoing Incendiary Illnesses: Relentless Resistant Enactment:

Ongoing incendiary illnesses result from delayed or dysregulated safe reactions. Conditions like fiery entrail sickness (IBD), psoriasis, and rheumatoid joint pain include tenacious aggravation, tissue harm, and disabled goal systems. Understanding the immunopathology of constant infections is essential for creating designated treatments.

5. Immunotherapy: Saddling the Invulnerable Framework for Treatment

5.1 Disease Immunotherapy: Releasing the Invulnerable Reaction against Growths:

Disease immunotherapy use the invulnerable framework's inherent capacity to perceive and take out malignant growth cells. Procedures incorporate resistant designated spot inhibitors, which release Lymphocytes to target growths, and assenting cell treatments, where designed safe cells are implanted to upgrade antitumor reactions. Immunotherapy has altered disease treatment, accomplishing sturdy reactions in different malignancies.

5.2 Antibodies: Taking action Framework for Safeguard:

Antibodies are useful assets that train the invulnerable framework to perceive and recall explicit microbes. They prompt an insusceptible reaction without causing infection, prompting the development of memory cells. Immunization has been instrumental in forestalling irresistible illnesses, with worldwide effect apparent in the destruction of smallpox and the control of sicknesses like polio and measles.

5.3 Immunomodulatory Treatments: Adjusting Resistant Reactions:

Immunomodulatory treatments intend to balance the safe framework, either improving or smothering explicit reactions. Corticosteroids and nonsteroidal mitigating drugs (NSAIDs) are instances of immunosuppressive specialists used to oversee provocative circumstances. Biologics, for example, monoclonal antibodies, target explicit resistant pathways in conditions like rheumatoid joint pain and psoriasis.

6. Future Outskirts: Progressions in Immunology

6.1 Immunometabolism: Meeting Pathways of Invulnerable and Metabolic Reactions:

Immunometabolism investigates the many-sided associations among invulnerable and metabolic pathways. Cell digestion impacts safe cell capability, and resistant reactions, thus, influence metabolic cycles. Understanding immunometabolism gives experiences into sicknesses like heftiness, diabetes, and fiery circumstances, preparing for novel restorative methodologies.

6.2 Single-Cell Immunology: Profiling Invulnerable Variety:

Progressions in single-cell advancements permit scientists to investigate the heterogeneity of safe cell populaces at uncommon goal. Single-cell immunology gives experiences into cell variety, useful states, and connections inside the safe framework. This innovation has suggestions for understanding illness pathogenesis and fitting accuracy medication draws near.

6.3 CRISPR-Cas9 and Quality Altering: Accuracy Instruments for Immunology:

The CRISPR-Cas9 quality altering framework has changed immunology research by empowering exact adjustment of qualities in resistant cells. This innovation works with the investigation of quality capability, the advancement of cell treatments, and the expected revision of hereditary deformities basic immunodeficiencies. CRISPR-based approaches hold guarantee for propelling immunotherapy and customized medication.

6.4 Microbiome and Immunology: Investigating the Stomach Resistant Hub:

The stomach microbiome, a different local area of microorganisms in the gastrointestinal system, significantly impacts safe reactions. The stomach safe pivot assumes an essential part in keeping up with resistant homeostasis and answering microorganisms. Understanding the transaction between the microbiome and the invulnerable framework has suggestions for immune system illnesses, sensitivities, and irresistible infections.

6.1 Analyzing the immune system of angelfish and its connection to human immunity.

Understanding the resistant arrangement of different organic entities gives important experiences into the developmental powers molding invulnerable reactions. The angelfish (Pterophyllum spp.), known for its dynamic tones and effortless appearance, fills in as a fascinating model organic entity for immunological examinations. This investigation digs into the exceptional parts of the angelfish resistant framework, attracting equals to human insusceptibility and disentangling the developmental variations that have formed these significant guard systems.

2. The Angelfish Resistant Framework: An Outline

2.1 Natural Resistance in Angelfish: Quick First-Line Safeguards:

The inborn insusceptible framework, a central part of resistant reactions, fills in as the quick first line of safeguard against microbes. Angelfish show intrinsic insusceptible components, including actual hindrances, antimicrobial peptides, and phagocytic cells. The skin and mucous layers go about as actual boundaries, forestalling microbe passage, while antimicrobial peptides assume a part in direct microorganism obliteration. Phagocytic cells, like macrophages and neutrophils, add to the engulfment and end of trespassers.

2.2 Versatile Resistance in Angelfish: Customized Safeguards and Immunological Memory:

Versatile invulnerability, a more particular arm of the resistant framework, gives custom fitted guards against explicit microorganisms. Angelfish have components of versatile invulnerability, including T and B lymphocytes. Lymphocytes, closely resembling their human partners, assume a part in cell safe reactions, perceiving and dispensing with contaminated or variant cells. B cells produce antibodies, adding to the humoral insusceptible reaction. The presence of versatile resistance in angelfish recommends transformative protection of these basic safeguard components.

2.3 Immunoglobulins in Angelfish: Antibodies as Atomic Watchmen:

Immunoglobulins, or antibodies, are vital parts of versatile invulnerability. Angelfish produce immunoglobulins, exhibiting the presence of immunizer interceded reactions. Antibodies in angelfish probably capability comparably to those in people, perceiving and killing microorganisms. Exploring the variety and particularity of angelfish immunoglobulins gives experiences into the advancement of neutralizer based invulnerable reactions.

3. Transformative Contemplations: Equals with Human Invulnerability

3.1 Hereditary Homology: Disentangling Transformative Associations:

Hereditary homology among angelfish and people offers an establishment for investigating transformative associations in resistant framework improvement. Relative genomics permits specialists to distinguish shared qualities and pathways, unwinding the sub-atomic premise of resistant reactions. Concentrating on rationed hereditary components reveals insight into the developmental directions that have prompted the variety of safe frameworks saw across species.

3.2 Phylogenetic Experiences: Following Transformative Ancestries:

Phylogenetic examination, which analyzes the transformative connections between species, helps with following the rise of safe framework highlights. Angelfish have a place with the request Perciformes, and a phylogenetic methodology helps place them inside the more extensive setting of fish development. Examinations with other fish species and vertebrates add to understanding the transformative variations that have formed resistant framework variety.

3.3 Transformative Tensions: Microorganisms as Drivers of Resistant Advancement:

The co-transformative weapons contest among hosts and microorganisms applies specific tensions on invulnerable frameworks. Angelfish, occupying oceanic conditions wealthy in assorted microorganisms, possible face one of a kind pathogenic difficulties. Investigating the transformations in angelfish resistant reactions gives experiences into the specific tensions that have impacted the development of safe techniques, offering matches with the human insusceptible framework.

4. Safe Difficulties in Oceanic Conditions

4.1 Oceanic Microorganisms: Exploring a Microbial Scene:

Oceanic conditions harbor a bunch of microorganisms, including microbes, infections, and parasites. Angelfish, as occupants of freshwater biological systems, experience a different exhibit of oceanic microorganisms. Their insusceptible framework should be capable at perceiving and answering these microbial dangers. Researching the systems utilized by angelfish to battle sea-going microorganisms improves how we might interpret safe transformations in conditions wealthy in microbial variety.

4.2 Mucosal Insusceptibility: Guarding at the Connection point:

The mucosal surfaces of the gastrointestinal and respiratory plots act as basic connection points between the angelfish and its current circumstance. Mucosal resistance assumes an imperative part in forestalling microorganism section and keeping up with microbial equilibrium. The angelfish mucosal resistant framework, including mucosal-related lymphoid tissues (MALT), adds to protection instruments at these points of interaction. Similar investigation with human mucosal insusceptibility offers experiences into shared systems for safeguarding mucosal surfaces.

4.3 Effect of Natural Changes: Resistant Reactions in Fluctuating Circumstances:

Sea-going conditions are vulnerable to variances in temperature, water quality, and microbial sythesis. Angelfish, as ectothermic organic entities, may encounter adjustments in resistant reactions because of ecological changes. Researching the effect of ecological factors on angelfish invulnerable capability gives a unique viewpoint on how the resistant framework adjusts to different circumstances, with suggestions for figuring out insusceptible strength.

5. Infection Models and Human Wellbeing

5.1 Angelfish as Infection Models: Uncovering Systems of Safe Reaction:

The angelfish's helplessness to specific microbes makes it an important model for concentrating on irresistible illnesses. Scientists can use angelfish as a stage to research have microbe collaborations, invulnerable reactions, and the sub-atomic premise of infection movement. Sickness models in angelfish add to how we might interpret irresistible illnesses and proposition bits of knowledge into likely remedial procedures.

5.2 Zebrafish as a Near Model: Extending the Extent of Exploration:

While angelfish gives important bits of knowledge, zebrafish (Danio rerio), a firmly related animal types, is a deeply grounded model life form with a completely sequenced genome. Near investigations among angelfish and zebrafish upgrade the profundity of exploration, utilizing the hereditary and trial benefits of the two species. Zebrafish, with its straightforward incipient organisms and hereditary manipulability, offers extra roads for concentrating on safe reactions.

5.3 Translational Ramifications for Human Wellbeing: Utilizing Developmental Experiences:

The equals among angelfish and human invulnerable frameworks have translational ramifications for human wellbeing. Understanding the preserved components in safe reactions permits scientists to extrapolate discoveries from angelfish studies to human settings.

Experiences into developmental transformations might illuminate procedures for improving human insusceptible reactions, creating antibodies, and overseeing irresistible illnesses.

6. Difficulties and Future Headings

6.1 Immunogenetic Variety: Exploring the Intricacy of Insusceptible Qualities:

The immunogenetic variety inside angelfish populaces presents difficulties in unraveling the complexities of resistant reactions. Researching the fluctuation in resistant qualities and their useful ramifications requires a thorough methodology. Understanding how immunogenetic variety impacts illness powerlessness and invulnerable strength is a basic part of future exploration.

6.2 Ecological Effect on Invulnerable Capability: Disentangling Complex Communications:

The effect of natural variables on angelfish insusceptible capability stays a perplexing area of study. Explaining the communications between natural factors, like temperature, water quality, and microbial creation, and safe reactions requires incorporated research draws near. Unraveling the subtleties of these communications adds to an all encompassing comprehension of resistant variation in fluctuating conditions.

6.3 Relative Immunology: Coordinating Discoveries Across Species:

The field of relative immunology benefits from coordinating discoveries across assorted species. Relative investigations with other fish species, creatures of land and water, and vertebrates enhance the comprehension of resistant framework advancement. Cooperative endeavors that range numerous life forms upgrade the pertinence of exploration discoveries and add to a more thorough information base.

6.2 Insights into inflammatory processes and autoimmune disorders.

Irritation is a key and developmentally moderated reaction that assumes an essential part in the body's guard components. Nonetheless, when this complex interaction becomes dysregulated, it can prompt a range of issues, including immune system sicknesses. This investigation digs into the nuanced parts of incendiary cycles, the invulnerable framework's job, and the complicated instruments that underlie immune system problems.

2. The Elements of Irritation

2.1 Intense Irritation: A Quick and Facilitated Reaction:

Intense irritation is the body's quick and early reaction to injury or contamination. It is an exceptionally coordinated process including different cells, flagging particles, and vascular changes. The cardinal indications of intense aggravation — redness, intensity, enlarging, and torment — mirror the unique exchange between safe cells, veins, and tissues. Neutrophils, the specialists on call, relocate to the site of injury or disease to wipe out microbes and cell garbage.

2.2 Constant Irritation: Delayed Initiation and Tissue Harm:

Constant irritation, interestingly, is an industrious and drawn out safe reaction that can prompt tissue harm. It frequently emerges when the intense provocative cycle isn't settled as expected or when the resistant framework is dysregulated. In constant aggravation, safe cells like macrophages, lymphocytes, and fibroblasts add to tissue redesigning and fix. Nonetheless, delayed actuation can bring about blow-back to sound tissues.

2.3 Fiery Middle people: Cytokines and Chemokines in real life:

Fiery middle people, including cytokines and chemokines, assume key parts in coordinating safe reactions. Cytokines like interleukins (IL) and growth rot factor (TNF) balance the force and length of aggravation. Chemokines guide invulnerable cells to explicit locales, advancing their movement and collaboration. The equilibrium of these middle people is basic for a very much managed invulnerable reaction and tissue homeostasis.

3. Immune system Problems: When Self-Resistance Separates

3.1 Autoimmunity Characterized: The Safe Framework Betrayed Itself:

Immune system problems happen when the resistant framework, which is intended to perceive and wipe out unfamiliar intruders, erroneously focuses on the body's own tissues. This breakdown of self-resistance brings about safe reactions coordinated against ordinary cells and tissues. Immune system infections are different, influencing different organs and frameworks, and they frequently include an intricate exchange of hereditary and ecological variables.

3.2 Normal Immune system Sicknesses: A Multi-layered Range:

A few immune system sicknesses influence various organs and tissues, each with its interesting clinical indications. Rheumatoid joint pain focuses on the joints, causing irritation and joint harm.

Foundational lupus erythematosus (SLE) is a fundamental immune system illness influencing different organs, with side effects going from skin rashes to kidney brokenness. Type 1 diabetes includes the safe framework going after insulin-delivering cells in the pancreas.

3.3 Systems of Autoimmunity: Unwinding the Triggers:

The systems hidden autoimmunity are perplexing and multifactorial. Hereditary inclination assumes a urgent part, with specific qualities related with an expanded gamble of immune system illnesses. Natural elements, like diseases, chemicals, and openness to specific substances, can set off or fuel immune system reactions. Sub-atomic mimicry, where microbial antigens look like self-antigens, can prompt resistant disarray and independent assaults.

4. Fiery Cycles in Immune system Issues

4.1 Ongoing Irritation in Autoimmunity: A Consistent theme:

Ongoing irritation is a sign of numerous immune system problems and adds to tissue harm and organ brokenness. In conditions like rheumatoid joint inflammation, the persistent irritation in the joints prompts the obliteration of ligament and bone. Fiery penetrates in organs, like the kidneys in lupus or the pancreas in diabetes, describe the movement of immune system illnesses.

4.2 Job of Resistant Cells in Autoimmunity: Dysregulation and Assault:

Resistant cells, including T lymphocytes and B lymphocytes, are key members in immune system reactions. In immune system issues, White blood cells might perceive self-antigens as unfamiliar, setting off a resistant reaction. B cells can deliver autoantibodies that target self-tissues, adding to irritation and tissue harm. The dysregulation of safe cell capability in autoimmunity highlights the intricacy of these problems.

4.3 Cytokines and Chemokines in Immune system Aggravation: Enhancing the Reaction:

The dysregulated creation of cytokines and chemokines is a trademark component of immune system irritation. Favorable to provocative cytokines, like TNF and IL-6, add to the propagation of aggravation and tissue harm. Chemokines draw in resistant cells to the kindled destinations, making a criticism circle that supports the immune system reaction. Focusing on these go betweens has turned into a restorative procedure in overseeing immune system problems.

5. Hereditary qualities, Ecological Triggers, and Autoimmunity

5.1 Hereditary Variables: The Diagram of Immune system Inclination:

Hereditary factors fundamentally impact the helplessness to immune system illnesses. Certain qualities related with the safe framework, like those in the human leukocyte antigen (HLA) district, assume an essential part. Varieties in these qualities can add to a singular's inclination to immune system issues. Be that as it may, hereditary qualities alone isn't adequate, and natural elements assume a basic part in setting off autoimmunity.

5.2 Natural Triggers: Exposing Immune system Reactions:

Ecological variables go about as triggers that can expose idle immune system inclinations. Contaminations, especially popular diseases, have been ensnared in the commencement or worsening of immune system illnesses. Hormonal changes, openness to specific medications or synthetic substances, and way of life variables may likewise impact the advancement of autoimmunity. Understanding the exchange among hereditary qualities and climate is pivotal for disentangling the intricacies of immune system problems.

5.3 The Cleanliness Speculation: Adjusting Invulnerable Enactment:

The cleanliness speculation suggests that decreased openness to contaminations and microbial upgrades in early life might add to the expanded predominance of immune system issues. An absence of early safe difficulties might bring about an imbalanced insusceptible framework, inclining people toward overstated reactions against self-antigens. This speculation features the unpredictable connection between the resistant framework, the microbiome, and natural elements.

6. Finding and Treatment Approaches

6.1 Demonstrative Difficulties: Exploring the Intricacy of Autoimmunity:

Diagnosing immune system problems can be trying because of the heterogeneity of side effects and the cross-over with other ailments. Research center tests, including autoantibody measures and imaging review, assume a pivotal part in affirming analyze. Nonetheless, a multidisciplinary approach including clinical assessment, clinical history, and coordinated effort between experts is in many cases fundamental for exact conclusion.

6.2 Immunosuppressive Treatments: Overseeing Aggravation and Autoimmunity:

Immunosuppressive treatments structure the foundation of overseeing immune system issues. Corticosteroids, which hose resistant reactions, are generally used to control irritation. Sickness altering antirheumatic drugs (DMARDs) and biologics target explicit safe pathways to regulate immune system reactions.

While these treatments are viable, they accompany possible aftereffects, stressing the requirement for customized treatment draws near.

6.3 Advances in Designated Treatments: Accuracy Medication in Autoimmunity:

Progresses in understanding the sub-atomic pathways engaged with autoimmunity have prompted the advancement of designated treatments. Biologics, including monoclonal antibodies, target explicit cytokines or safe cells associated with immune system irritation. Janus kinase (JAK) inhibitors and other little particles offer extra choices for accuracy medication, permitting more customized ways to deal with overseeing immune system issues.

7. Future Bearings and Difficulties

7.1 Customized Medication in Autoimmunity: Fitting Treatment Techniques:

The time of customized medication holds guarantee for fitting treatment systems in immune system issues. Distinguishing explicit hereditary and sub-atomic marks related with individual cases might direct the determination of designated treatments. Accuracy approaches expect to upgrade treatment results while limiting unfriendly impacts, denoting a change in outlook in the administration of immune system sicknesses.

7.2 Unwinding the Microbiome-Resistant Pivot: Investigating New Wildernesses:

The exchange between the microbiome and the resistant framework is a prospering area of exploration in autoimmunity. The stomach microbiome, specifically, impacts safe reactions and may assume a vital part in immune system guideline. Understanding how the microbiome shapes safe homeostasis offers novel roads for remedial intercessions and preventive systems.

7.3 Patient Instruction and Strengthening: Exploring the Intricacies:

Patient instruction and strengthening are fundamental parts of overseeing immune system issues. Improving mindfulness about the idea of these circumstances, treatment choices, and way of life factors engages people to partake in their medical services effectively. Support gatherings and patient backing associations add to an aggregate exertion in bringing issues to light and cultivating a feeling of local area among those impacted via immune system sicknesses.

6.3Potential advancements in immunotherapy.

Immunotherapy, a noteworthy methodology bridling the body's resistant framework to battle illnesses, has arisen as a groundbreaking power in clinical science. From malignant growth to immune system problems, the possible headways in immunotherapy hold guarantee for changing the scene of sickness therapy.

This investigation digs into the state of the art advancements, challenges, and the extraordinary effect of immunotherapy on assorted clinical fronts.

2. Immunotherapy in Disease Treatment

2.1 Resistant Designated spot Inhibitors: Upgrading Lymphocyte Reactions:

Resistant designated spot inhibitors, a foundation of malignant growth immunotherapy, release the force of Lymphocytes against disease cells. By obstructing inhibitory pathways like PD-1/PD-L1 or CTLA-4, these inhibitors eliminate the brakes on the safe framework, permitting Lymphocytes to perceive and go after malignant growth cells. Progressions in recognizing novel designated spots and mix treatments plan to improve reaction rates and expand the materialness of designated spot inhibitors across various malignant growth types.

2.2 Vehicle Immune system microorganism Treatment: Accuracy Focusing of Malignant growth Cells:

Fanciful Antigen Receptor Immune system microorganism (Vehicle T) treatment includes hereditarily changing a patient's Lymphocytes to communicate receptors focusing on unambiguous disease antigens. This customized approach has shown noteworthy achievement, especially in hematological malignancies. Continuous exploration centers around refining Vehicle White blood cell treatment, tending to difficulties like askew impacts and growing its viability to strong cancers. Cutting edge Vehicle T plans and progressions in assembling methods mean to make this treatment more available and powerful.

2.3 Oncolytic Infections: Taking advantage of Viral Specialists against Malignant growth:

Oncolytic infections are designed or normally happening infections that specifically contaminate and obliterate malignant growth cells. These infections straightforwardly lyse malignant growth cells as well as animate the insusceptible framework's enemy of cancer reactions. Scientists are investigating the capability of oncolytic infections in blend with different immunotherapies to make synergistic impacts. Fitting viral specialists for explicit malignant growth types and improving conveyance techniques are areas of dynamic examination.

3. Immunotherapy in Immune system Problems

3.1 Insusceptible Adjustment: Adjusting Hyperactive Reactions:

In immune system problems where the safe framework goes after the body's own tissues, resistant regulation is a key technique. Headways in immunotherapy for immune system illnesses include specifically hosing hyperactive safe reactions.

Biologics focusing on unambiguous cytokines, cell subsets, or flagging pathways expect to adjust safe movement without worldwide immunosuppression. Customized approaches that consider the heterogeneity of immune system issues and individual patient reactions are turning out to be progressively significant.

3.2 Resistance Acceptance: Diverting Insusceptible Reactions:

Resistance acceptance procedures try to reinstruct the invulnerable framework to endure self-antigens, lessening the penchant for immune system assaults. Exploratory methodologies include uncovering the safe framework to controlled dosages of self-antigens, advancing the improvement of administrative White blood cells that smother distorted invulnerable reactions. Progressions in understanding the components of resistance enlistment and enhancing treatment regimens hold guarantee for more secure and more successful treatments.

4. Immunotherapy in Irresistible Illnesses

4.1 Immunizations: Past Anticipation to Remedial Mediation:

Customary immunizations mean to forestall contaminations, yet the capability of immunotherapy stretches out to helpful antibodies intended to treat existing diseases. Helpful immunizations invigorate the insusceptible framework to perceive and go after microorganisms, offering an original way to deal with fighting persistent diseases like HIV and hepatitis. Headways in immunization plan, adjuvant turn of events, and customized antibody systems add to the advancing scene of helpful inoculation.

4.2 Invulnerable Enhancers: Helping the Body's Guards:

Safe enhancers, otherwise called immunomodulators, are substances that invigorate the invulnerable framework to upgrade its defensive reactions. These enhancers can be utilized as assistants to ordinary antimicrobial treatments, supporting the body's capacity to battle contaminations. Research centers around distinguishing explicit focuses for resistant upgrade and creating methodologies to adjust insusceptible reactions for ideal viability.

5. Difficulties and Future Bearings

5.1 Beating Opposition in Disease Immunotherapy:

While disease immunotherapy has shown striking achievement, obstruction stays a test. Growths can foster systems to avoid safe acknowledgment or stifle resistant reactions inside the growth microenvironment. Conquering obstruction includes unwinding the complexities of cancer invulnerable associations, creating mix treatments, and recognizing prescient biomarkers to direct therapy choices.

5.2 Personalization and Biomarkers: Fitting Immunotherapy Approaches:

The mission for customized immunotherapy includes distinguishing biomarkers that anticipate treatment reactions and guide restorative decisions. Progresses in genomics, transcriptomics, and safe profiling add to the disclosure of prescient biomarkers. Fitting immunotherapy approaches in view of individual patient qualities expects to upgrade results, limit aftereffects, and improve the general viability of treatment.

5.3 Overseeing Immune system Secondary effects: Adjusting Adequacy and Security:

Immunotherapy, while groundbreaking, can be related with immune system secondary effects known as insusceptible related unfavorable occasions (irAEs). Finding some kind of harmony between amplifying restorative adequacy and it is essential to limit irAEs. Progressing research centers around figuring out the systems of irAEs, creating techniques for early location, and refining treatment calculations to really deal with these secondary effects.

5.4 Widening Access and Moderateness: Conquering Hindrances to Treatment:

In spite of the progress of immunotherapy, openness and moderateness remain difficulties. Significant expenses, calculated intricacies, and variations in medical services access limit the far and wide reception of immunotherapies. Research endeavors intend to foster savvy fabricating processes, smooth out calculated viewpoints, and execute techniques to guarantee impartial admittance to these extraordinary treatments.

Chapter 7

Metabolic Pathways and Diseases

Metabolic pathways are the perplexing organizations of biochemical responses that happen inside cells to create energy, combine biomolecules, and keep up with cell capabilities. The fragile equilibrium of these pathways is fundamental for cell homeostasis and in general organismal wellbeing. This investigation digs into the intricacy of metabolic pathways, their guideline, and the vital job they play in the turn of events and movement of different illnesses.

2. Outline of Metabolic Pathways

2.1 Glycolysis: The Focal Center of Energy Creation:

Glycolysis, the underlying move toward cell energy digestion, includes the breakdown of glucose into pyruvate. This cycle happens in the cytoplasm and yields ATP and NADH, giving the cell fundamental energy cash. Glycolysis fills in as a focal center point, taking care of into different downstream metabolic pathways.

2.2 Citrus extract Cycle (TCA Cycle): Extricating Energy from Pyruvate:

The TCA cycle, happening in the mitochondria, further oxidizes pyruvate to create electron transporters (NADH and FADH2) and GTP. These electron transporters assume a critical part in oxidative phosphorylation, the cycle liable for most of ATP creation. The TCA cycle associates different metabolic pathways, adding to energy age and the amalgamation of forerunner particles.

2.3 Oxidative Phosphorylation: ATP Combination in the Mitochondria:

Oxidative phosphorylation, completed in the mitochondrial electron transport chain, bridles the energy put away in electron transporters to produce ATP. Electrons move through protein buildings, making a proton inclination across the inward mitochondrial film. The resulting stream of protons back into the mitochondrial network powers ATP blend, featuring the interconnectedness of cell energy creation.

2.4 Lipid Digestion: Fuel Stockpiling and Cell Layers:

Lipid digestion includes the amalgamation and breakdown of fats, critical for energy capacity and film structure. Fatty oils, framed from glycerol and unsaturated fats, act as the essential stockpiling type of energy in fat tissue. Lipid digestion additionally adds to the age of flagging atoms and the support of cell film respectability.

2.5 Amino Corrosive Digestion: Building Blocks and Energy Sources:

Amino corrosive digestion incorporates the combination and debasement of amino acids, the structure blocks of proteins. Past their part in protein combination, amino acids add to energy creation through their change to intermediates in metabolic pathways. The crosstalk between amino corrosive digestion and different pathways is essential for cell capability and homeostasis.

3. Metabolic Guideline: Difficult exercise for Cell Homeostasis

3.1 Hormonal Guideline: Planning Reactions to Supplement Accessibility:

Chemicals, like insulin and glucagon, assume a focal part in planning metabolic reactions to supplement accessibility. Insulin advances glucose take-up and capacity, restraining gluconeogenesis and glycogenolysis. Interestingly, glucagon flags the arrival of glucose from glycogen stores and invigorates gluconeogenesis. Hormonal guideline guarantees a unique harmony between energy capacity and use.

3.2 Enzymatic Control: Adjusting Metabolic Pathways:

Proteins go about as impetuses in metabolic responses, and their movement is firmly managed to keep up with cell homeostasis. Allosteric guideline, criticism hindrance, and post-translational alterations tweak protein action in light of changing cell conditions. Enzymatic control permits cells to adjust to fluctuating energy requests and substrate accessibility.

3.3 Cell Flagging Pathways: Mix of Metabolic Signs:

Cell flagging pathways, including AMP-initiated protein kinase (AMPK) and mammalian objective of rapamycin (mTOR), coordinate metabolic signs to direct energy balance. AMPK, enacted during energy stress, advances catabolic cycles and represses anabolic pathways. mTOR, delicate to supplement accessibility, animates cell development and protein union when conditions are positive.

3.4 Epigenetic Guideline: Metabolic Effects on Quality Articulation:

Epigenetic changes, for example, DNA methylation and histone acetylation, assume a part in metabolic guideline by impacting quality articulation. Metabolites created in different pathways can act as cofactors for chemicals engaged with epigenetic adjustments. This perplexing exchange associates cell digestion with the guideline of key qualities associated with energy equilibrium and homeostasis.

4. Metabolic Infections: Interruptions to Cell Amicability

4.1 Sort 2 Diabetes: Dysregulation of Glucose Homeostasis:

Type 2 diabetes is portrayed by insulin opposition and debilitated glucose homeostasis. Dysregulation of metabolic pathways, especially in insulin-responsive tissues like muscle and fat tissue, prompts raised blood glucose levels. The interaction of hereditary and ecological variables adds to the advancement of insulin obstruction, connecting metabolic brokenness to the pathogenesis of diabetes.

4.2 Weight: Unevenness in Energy Stockpiling and Use:

Weight results from an unevenness between energy admission and use, prompting the unreasonable aggregation of fat tissue. Dysregulation of lipid digestion, combined with persistent second rate irritation, adds to insulin opposition and metabolic difficulties. Fat tissue-determined signals, for example, adipokines, assume a part in foundational metabolic homeostasis.

4.3 Cardiovascular Sicknesses: Lipid Digestion and Atherosclerosis:

Disturbances in lipid digestion add to the advancement of cardiovascular sicknesses, including atherosclerosis. Raised degrees of low-thickness lipoprotein (LDL) cholesterol, got from uneven characters in lipid digestion, can collect in blood vessel walls, setting off irritation and plaque arrangement. The mind boggling connection between lipid digestion and cardiovascular wellbeing highlights the significance of keeping up with metabolic concordance.

4.4 Non-Alcoholic Greasy Liver Illness (NAFLD): Hepatic Lipid Gathering:

NAFLD addresses a range of liver circumstances portrayed by unreasonable lipid collection in hepatocytes. Dysregulation of lipid digestion, insulin obstruction, and irritation add to the movement of NAFLD. The infection's relationship with weight and metabolic condition features the interconnectedness of fundamental digestion and liver wellbeing.

5. Restorative Methodologies and Future Headings

5.1 Focusing on Metabolic Pathways in Illness Treatment:

Helpful procedures focusing on metabolic pathways offer imaginative methodologies for overseeing metabolic illnesses. In type 2 diabetes, meds mean to further develop insulin responsiveness and direct glucose homeostasis. Drugs focusing on lipid digestion, for example, statins, assume a part in overseeing cardiovascular sicknesses. Continuous exploration investigates novel mixtures and accuracy medication ways to deal with improve remedial results.

5.2 Accuracy Medication and Customized Treatments: Fitting Mediations:

Accuracy medication, utilizing progresses in genomics and sub-atomic profiling, holds guarantee for fitting mediations to individual metabolic profiles. Customized treatments might include focusing on unambiguous hereditary variations, advancing medication regimens in view of metabolic attributes, and carrying out way of life mediations custom-made to a singular's novel metabolic necessities.

5.3 Nutrigenomics: Collaborations Among Diet and Hereditary qualities:

Nutrigenomics investigates the communications among diet and a person's hereditary cosmetics. Understanding how hereditary varieties impact reactions to dietary parts takes into consideration customized nourishment proposals. Nutrigenomic approaches think about the effect of explicit supplements on metabolic pathways, giving experiences into preventive procedures and remedial intercessions.

5.4 Metabolomics: Disentangling Metabolic Marks:

Metabolomics, the extensive examination of metabolites in natural examples, offers bits of knowledge into metabolic marks related with sickness states. By profiling metabolites, scientists can distinguish biomarkers for early infection recognition, screen treatment reactions, and unwind the unpredictable connections between metabolic dysregulation and different obsessive circumstances.

7.1Investigating metabolic pathways in angelfish and their relevance to human metabolism.

Metabolic pathways, the mind boggling organizations of biochemical responses administering energy creation and biomolecule union, are central to the working, everything being equal. While human digestion has been broadly considered, investigating metabolic pathways in assorted model life forms gives special experiences. This investigation digs into the examination of metabolic pathways in angelfish (Pterophyllum spp.) and expects to clarify their significance to human digestion.

2. Angelfish as a Model Organic entity

2.1 Qualities of Angelfish:

Angelfish, known for their unmistakable appearance and agile developments, are freshwater cichlids local to South America. Having a place with the Pterophyllum variety, they occupy sluggish waterways and display complex social ways of behaving. Angelfish definitely stand out enough to be noticed as a model creature in biomedical examination because of their hereditary manageability, simplicity of upkeep in research facility settings, and physiological likenesses to different vertebrates.

2.2 Benefits of Angelfish as a Model:

Angelfish offer a few benefits as a model creature for metabolic investigations. Their generally little size works with savvy support in research center settings, and their regenerative examples consider the age of hereditarily controlled populaces. Moreover, the accessibility of genomic instruments for angelfish empowers scientists to investigate their sub-atomic and hereditary qualities, making them a significant framework for researching metabolic pathways.

3. Metabolic Pathways in Angelfish

3.1 Glycolysis and Energy Creation:

Glycolysis, the general pathway for glucose breakdown, assumes a significant part in energy creation. In angelfish, as in people, glycolysis happens in the cytoplasm, producing ATP and pyruvate from glucose. Exploring glycolytic proteins and administrative systems in angelfish gives experiences into the preservation of energy creation pathways across species.

3.2 Citrus extract Cycle (TCA Cycle) and Mitochondrial Capability:

The TCA cycle, happening in the mitochondria, is vital for extricating energy from metabolic intermediates. Angelfish share likenesses in TCA cycle chemicals and mitochondrial capability with different vertebrates. Investigating the guideline of the TCA cycle in angelfish adds to how we might interpret mitochondrial digestion and its suggestions for cell energy balance.

3.3 Lipid Digestion and Capacity:

Lipid digestion, including the combination and breakdown of fats, is necessary to energy capacity and film structure. Angelfish, in the same way as other vertebrates, store energy as fatty substances. Researching the chemicals engaged with lipid digestion in angelfish gives experiences into the systems administering energy stockpiling and use, with possible ramifications for understanding corpulence related messes in people.

3.4 Amino Corrosive Digestion and Protein Blend:

Amino corrosive digestion, enveloping the union and debasement of amino acids, is imperative for protein combination and energy creation. Angelfish use amino acids as building blocks for protein amalgamation, like different vertebrates. Concentrating on the guideline of amino corrosive digestion in angelfish reveals insight into the monitored systems administering protein combination and cell homeostasis.

4. Near Examination: Pertinence to Human Digestion

4.1 Transformative Preservation of Metabolic Pathways:

Transformative preservation of metabolic pathways highlights the similitudes in essential biochemical cycles across species. Similar examination uncovers shared enzymatic exercises, administrative components, and metabolic intermediates among angelfish and people. This protection gives a premise to extrapolating discoveries from angelfish studies to improve how we might interpret human digestion.

4.2 Experiences into Metabolic Variations:

While metabolic pathways are rationed, species-explicit variations likewise exist. Exploring metabolic variations in angelfish, like reactions to shifting ecological circumstances or dietary changes, offers bits of knowledge into the adaptability of metabolic organizations. Understanding these transformations upgrades our insight into how metabolic pathways develop to meet the exceptional necessities of various creatures.

4.3 Metabolic Problems and Sickness Demonstrating:

Metabolic issues, going from diabetes to lipid digestion related illnesses, present huge wellbeing challenges in people. Angelfish, as a model life form, can be instrumental in concentrating on the sub-atomic premise of metabolic issues. By prompting explicit hereditary or ecological bothers, analysts can explore the turn of events and movement of metabolic illnesses, giving significant experiences to likely remedial mediations in people.

4.4 Medication Disclosure and Translational Exploration:

Metabolic pathways are ideal objectives for drug improvement, particularly with regards to metabolic infections. Angelfish, with their preserved metabolic pathways, act as an important model for drug revelation. Concentrating on the impacts of drug intensifies on angelfish digestion can give starter information to translational exploration, directing resulting examinations in mammalian frameworks.

5. Mechanical Advances in Angelfish Metabolomics

5.1 Metabolomics Approaches:

Metabolomics, the exhaustive examination of little particles in organic examples, has turned into an amazing asset for concentrating on metabolic pathways. In angelfish research, propels in metabolomics innovations empower scientists to profile and evaluate a great many metabolites. This frameworks level methodology works with the ID of metabolic marks and modifications related with hereditary or ecological irritations.

5.2 Genomic Devices and CRISPR/Cas9 Innovation:

Genomic devices, including clarified genomes and hereditary data sets, give an establishment to grasping the sub-atomic premise of metabolic cycles in angelfish. The appearance of CRISPR/Cas9 innovation considers exact genome altering, empowering designated alterations to concentrate on the capability of explicit qualities engaged with metabolic pathways. These instruments upgrade the accuracy and effectiveness of angelfish metabolic examination.

5.3 Bioinformatics and Information Coordination:

As the volume of information produced from metabolomics and genomics concentrates on expands, bioinformatics and information reconciliation become basic. Logical instruments and bioinformatic approaches permit specialists to decipher complex datasets, distinguish key metabolic pathways, and reveal expected administrative hubs. Incorporating information from various sources upgrades the profundity of how we might interpret angelfish digestion.

6. Future Bearings and Difficulties

6.1 Disentangling Metabolic Adaptability:

Examining the metabolic adaptability of angelfish because of changing natural circumstances presents a promising road for future examination. Understanding how angelfish adjust their metabolic pathways to changes in temperature, diet, or different variables gives important data on the versatility of metabolic organizations.

6.2 Investigating Metabolic Associations:

Metabolic pathways don't work in segregation; they cooperate with one another to keep up with cell homeostasis. Future exploration in angelfish digestion ought to investigate these connections, revealing the crosstalk between various pathways and the integrative components that arrange metabolic reactions to different improvements.

6.3 Difficulties in Deciphering Discoveries:

Deciphering discoveries from angelfish digestion to human importance accompanies difficulties. While the basic metabolic pathways are rationed, species-explicit contrasts exist. Analysts should cautiously consider these distinctions and lead extra examinations in mammalian models to approve the appropriateness of angelfish discoveries to human digestion.

7.2 Implications for understanding and treating metabolic disorders.

Metabolic issues, enveloping a range of conditions from diabetes to lipid digestion related illnesses, present huge difficulties to worldwide wellbeing. Understanding the hidden atomic systems and creating compelling medicines for these problems are vital. This investigation dives into the ramifications of angelfish research for propelling comprehension we might interpret metabolic issues and the potential it holds for imaginative remedial intercessions.

2. Uncovering Sub-atomic Experiences through Angelfish Digestion

2.1 Preservation of Metabolic Pathways:

Angelfish, as model life forms, give a remarkable window into the protection of metabolic pathways across species. The key biochemical cycles overseeing energy creation, lipid digestion, and amino corrosive usage are surprisingly preserved among angelfish and people. Researching these pathways in angelfish offers bits of knowledge into the atomic premise of metabolic problems and their developmental starting points.

2.2 Species-Explicit Variations:

While metabolic pathways are preserved, species-explicit transformations exist to meet the extraordinary requirements of every living being. Angelfish, possessing assorted conditions right at home, display metabolic variations to fluctuating circumstances. Concentrating on these transformations improves how we might interpret how metabolic adaptability and flexibility add to in general wellbeing and may give hints to tending to metabolic problems in people.

3. Illustrations from Angelfish Metabolomics

3.1 Metabolomics Approaches:

Headways in metabolomics advancements empower exhaustive profiling of metabolites in angelfish, revealing insight into the many-sided trap of sub-atomic collaborations inside metabolic pathways. The ID of metabolic marks related with explicit circumstances or hereditary bothers gives an establishment to disentangling the intricacies of metabolic issues.

3.2 Biomarker Revelation:

Metabolomics in angelfish research adds to biomarker revelation, a significant part of understanding and diagnosing metabolic problems. Recognizing explicit metabolites or examples related with metabolic brokenness in angelfish gives potential biomarkers that might mean human examinations. Biomarker revelation works with early illness location and observing treatment reactions.

3.3 Natural Impacts on Digestion:

Angelfish digestion reflects transformations to ecological variables, like temperature and diet. Concentrating on the effect of these variables on metabolic pathways in angelfish gives experiences into how outer impacts add to metabolic wellbeing.

Understanding ecological effects on digestion is fundamental for tending to way of life related metabolic problems in people.

4. Demonstrating Metabolic Problems in Angelfish

4.1 Hereditary Control and Sickness Displaying:

The hereditary manageability of angelfish, combined with CRISPR/Cas9 innovation, permits analysts to prompt explicit hereditary changes. This capacity works with the production of angelfish models that mirror parts of human metabolic problems. By presenting hereditary adjustments related with illnesses like diabetes or weight, specialists can concentrate on the movement and basic instruments of these issues.

4.2 Bits of knowledge into Sickness Pathogenesis:

Angelfish models of metabolic issues offer bits of knowledge into illness pathogenesis at the sub-atomic level. Concentrating on the impacts of hereditary changes or ecological variables on metabolic pathways in angelfish explains the systems prompting insulin opposition, dyslipidemia, and different signs of metabolic issues. This information is urgent for creating designated mediations in people.

5. Translational Potential: From Angelfish to Human Wellbeing

5.1 Focusing on Shared Pathways:

The preserved idea of metabolic pathways among angelfish and people takes into consideration the extrapolation of discoveries from angelfish examination to human wellbeing. Disclosures connected with key chemicals, administrative hubs, or metabolic intermediates in angelfish digestion might illuminate the improvement regarding designated treatments for human metabolic problems. Shared pathways give shared belief to restorative mediations.

5.2 Accuracy Medication Approaches:

Accuracy medication, fitting medicines to individual qualities, is a promising road in the administration of metabolic problems. Experiences acquired from angelfish studies can add to the advancement of accuracy medication approaches by distinguishing hereditary variations or metabolic marks related with explicit reactions to mediations. This customized approach improves treatment viability and limits secondary effects.

5.3 Medication Disclosure and Screening:

Angelfish research assumes a part in drug disclosure for metabolic issues. By testing the impacts of drug intensifies on angelfish digestion, specialists can distinguish potential medication applicants and survey their viability. This underlying evaluating in angelfish models gives an establishment to ensuing examinations in mammalian frameworks, smoothing out the medication improvement process.

6. Difficulties and Contemplations in Interpretation

6.1 Species-Explicit Contrasts:

Regardless of the preservation of metabolic pathways, species-explicit contrasts exist among angelfish and people. Analysts should cautiously consider these distinctions while deciphering discoveries from angelfish studies to human significance. Approval in mammalian models is fundamental to affirm the relevance of mediations or restorative targets recognized in angelfish.

6.2 Intricacies of Infection Aggregates:

Demonstrating metabolic problems in angelfish may not completely repeat the intricacy of human illness aggregates. Human metabolic issues frequently include diverse associations between hereditary, ecological, and way of life factors. Analysts ought to perceive the limits of angelfish models and supplement their discoveries with information from assorted mammalian frameworks.

7. Future Bearings: Progressing Metabolic Exploration with Angelfish

7.1 Incorporation of Omics Advancements:

The combination of omics advancements, including genomics, metabolomics, and transcriptomics, upgrades the profundity of figuring out in angelfish digestion. Far reaching datasets produced through multi-omics approaches permit specialists to investigate the complicated organizations of sub-atomic cooperations administering metabolic pathways and their dysregulation in sickness states.

7.2 Investigation of Stomach Microbiota Impact:

The stomach microbiota assumes a pivotal part in forming host digestion. Future exploration in angelfish digestion ought to investigate the impact of stomach microbiota on metabolic pathways. Grasping the interchange between have digestion and the microbiome adds to a comprehensive comprehension of metabolic wellbeing and expected intercessions for metabolic issues.

7.3 Ecological and Nourishing Investigations:

Given the awareness of angelfish digestion to ecological factors and diet, future examinations ought to dive into the effect of explicit natural circumstances and wholesome creations on metabolic pathways. This examination can give important experiences into preventive methodologies and way of life intercessions for tending to metabolic issues.

7.3 Dietary and lifestyle recommendations based on angelfish research.

Diet and way of life decisions assume a crucial part in molding metabolic wellbeing, impacting energy balance, and relieving the gamble of metabolic problems. While human examinations have given significant bits of knowledge, investigating the dietary and way of life proposals in light of angelfish research offers an exceptional viewpoint. This investigation dives into the illustrations gained from angelfish studies, giving bits of knowledge into dietary and way of life intercessions that might have suggestions for human wellbeing.

2. Angelfish as a Model for Dietary and Way of life Exploration

2.1 Supplement Necessities and Metabolic Transformations:

Angelfish, occupying assorted conditions with shifting supplement accessibility, show metabolic transformations to various weight control plans. Concentrating on the supplement prerequisites and metabolic reactions of angelfish to explicit dietary parts gives bits of knowledge into how dietary decisions impact metabolic pathways. This information can direct proposals for advancing supplement admission in people.

2.2 Effect of Ecological Elements:

The ecological circumstances in angelfish natural surroundings, like water temperature and quality, impact their digestion and by and large wellbeing. Understanding what these elements mean for angelfish gives an establishment to thinking about natural impacts in dietary and way of life proposals for people. It underlines the interconnectedness of ecological elements with metabolic prosperity.

3. Dietary Suggestions from Angelfish Studies

3.1 Protein Admission and Development:

Angelfish research features the significance of protein in development and improvement. Satisfactory protein admission is essential for supporting muscle development, upkeep, and generally metabolic wellbeing. While angelfish have explicit protein prerequisites, the discoveries highlight the overall significance of consolidating excellent protein sources in human weight control plans to help ideal metabolic capability.

3.2 Lipid Digestion and Fundamental Unsaturated fats:

Lipid digestion is a critical concentration in angelfish studies, particularly concerning the job of fundamental unsaturated fats. Angelfish, similar to people, require explicit unsaturated fats for appropriate development and improvement. The accentuation on fundamental unsaturated fats in angelfish slims down recommends the significance of consolidating wellsprings of omega-3 and omega-6 unsaturated fats in human eating regimens for cardiovascular wellbeing and in general metabolic equilibrium.

3.3 Sugar Use and Energy Equilibrium:

Angelfish digestion gives experiences into starch usage to energy creation. The harmony among sugars and other macronutrients is significant for keeping up with energy balance and forestalling metabolic unsettling influences. Understanding how angelfish manage energy from starches illuminates suggestions for people on picking complex, fiber-rich carbs for supported energy.

3.4 Micronutrient Necessities:

Angelfish, similar to all creatures, have explicit prerequisites for fundamental nutrients and minerals. Examination into angelfish micronutrient prerequisites adds to how we might interpret the job these components play in digestion. Adjusting angelfish discoveries to human sustenance underlines the significance of a different and supplement rich eating regimen to address micronutrient issues and backing metabolic wellbeing.

4. Dietary Proposals for Human Wellbeing

4.1 Accentuating Entire Food sources:

Angelfish research advocates for the utilization of entire food varieties to meet assorted supplement needs. Entire food sources, including organic products, vegetables, entire grains, lean proteins, and solid fats, give a range of supplements that add to generally speaking wellbeing. Focusing on entire food sources lines up with angelfish studies and is a foundation of dietary proposals for human wellbeing.

4.2 Adjusting Macronutrients:

The harmony between macronutrients, as seen in angelfish studies, is fundamental for metabolic wellbeing. Dietary proposals for people underscore a reasonable admission of starches, proteins, and fats. Fitting macronutrient proportions to individual necessities and taking into account the nature of these supplements add to ideal metabolic capability and weight the executives.

4.3 Fundamental Unsaturated fats for Cardiovascular Wellbeing:

Angelfish research highlights the significance of fundamental unsaturated fats for cardiovascular wellbeing. Omega-3 and omega-6 unsaturated fats, tracked down in greasy fish, flaxseeds, and pecans, assume a crucial part in lessening the gamble of cardiovascular sicknesses. Dietary proposals urge consolidating these sources to help heart wellbeing and keep a positive lipid profile.

4.4 Protein Quality and Muscle Wellbeing:

Protein quality is pivotal for keeping up with muscle wellbeing, as featured by angelfish studies. Human dietary suggestions stress the significance of consuming top notch protein sources, like lean meats, dairy, vegetables, and plant-based proteins. Satisfactory protein consumption upholds muscle upkeep, invulnerable capability, and in general metabolic equilibrium.

4.5 Micronutrient-Rich Eating regimen:

A micronutrient-rich eating regimen is crucial for metabolic wellbeing, drawing matches with angelfish micronutrient prerequisites. Dietary proposals energize the utilization of different brilliant products of the soil to guarantee a variety of nutrients and minerals. Micronutrients assume fundamental parts in enzymatic responses, energy digestion, and generally physiological prosperity.

5. Way of life Proposals In view of Angelfish Bits of knowledge

5.1 Actual work and Metabolic Adaptability:

Angelfish, right at home, display metabolic adaptability affected by actual work. Human way of life suggestions draw motivation from this, underscoring ordinary active work for metabolic wellbeing. Practice improves insulin responsiveness, advances energy consumption, and adds to weight the executives, lining up with angelfish research on metabolic variations to development.

5.2 Ecological Contemplations:

Ecological factors fundamentally influence angelfish digestion, mirroring the significance of thinking about the climate in way of life proposals. For people, way of life decisions like openness to regular light, satisfactory rest, and stress the board add to by and large metabolic prosperity. Adjusting ecological impacts upholds circadian rhythms, hormonal equilibrium, and ideal metabolic capability.

5.3 Hydration and Metabolic Homeostasis:

Angelfish concentrates on feature the job of water quality in metabolic wellbeing. In human way of life suggestions, sufficient hydration is fundamental for keeping up with metabolic homeostasis.

Water upholds processing, supplement transport, and temperature guideline. Integrating adequate water consumption into everyday schedules is a basic yet effective direction for living for metabolic wellbeing.

5.4 Pressure The board and Hormonal Equilibrium:

Stress is a huge element influencing angelfish digestion, underscoring the significance of stress the board in way of life suggestions. Ongoing pressure can upset hormonal equilibrium and add to metabolic problems. Way of life mediations like care, reflection, and adequate rest assume a pivotal part in moderating pressure and supporting generally metabolic wellbeing.

6. Difficulties and Contemplations in Execution

6.1 Individual Changeability:

Individual changeability in light of dietary and way of life mediations is a test featured by angelfish research. People show assorted metabolic reactions in view of hereditary qualities, climate, and way of life. Fitting suggestions to individual necessities and taking into account factors, for example, hereditary qualities and stomach microbiota upgrades the adequacy of dietary and way of life intercessions.

6.2 Financial Elements:

Financial variables impact admittance to different and nutritious food sources, influencing the capacity to follow dietary suggestions. Angelfish studies don't confront financial difficulties, and making an interpretation of proposals to human populaces requires tending to differences in food access, training, and monetary assets. Executing methodologies for fair sustenance is fundamental for broad metabolic wellbeing.

7. Future Bearings: Coordinating Angelfish Bits of knowledge into General Wellbeing

7.1 Local area Based Exploration:

Local area based research drives can coordinate angelfish bits of knowledge into general wellbeing intercessions. Cooperative undertakings including specialists, medical services experts, and networks can investigate the commonsense use of dietary and way of life suggestions. Drawing in networks cultivates socially applicable methodologies and upgrades the reception of sound ways of behaving.

7.2 Longitudinal Examinations:

Longitudinal examinations following the impacts of dietary and way of life intercessions in light of angelfish bits of knowledge give important information on supported metabolic wellbeing.

Understanding the drawn out effect of proposals upholds the advancement of proof based rules for people and populaces. Longitudinal exploration adds to refining proposals after some time.

7.3 Worldwide Wellbeing Systems:

Making an interpretation of angelfish-roused suggestions into worldwide wellbeing systems requires an extensive methodology. Incorporating metabolic wellbeing into more extensive general wellbeing drives, tending to social determinants of wellbeing, and upholding for strategies that help solid ways of life add to an all encompassing and maintainable effect on metabolic prosperity around the world.

Chapter 8

Translational Challenges and Future Directions

Angelfish, as a model living being, has given significant bits of knowledge into different parts of science, including digestion, hereditary qualities, and cardio-vascular capability. The translational capability of angelfish exploration to human wellbeing is immense, holding guarantee for progressions in illness getting it, drug revelation, and helpful mediations. Be that as it may, overcoming any barrier from angelfish studies to significant applications in human wellbeing faces difficulties and requires an essential methodology. This investigation dives into the translational difficulties looked by angelfish exploration and layouts future bearings to amplify its effect on human wellbeing.

2. Translational Difficulties

2.1 Species-Explicit Contrasts:

One of the essential provokes in making an interpretation of angelfish research discoveries to people lies in species-explicit contrasts. While essential natural cycles are saved, the interesting attributes and developmental variations of angelfish may not impeccably reflect human physiology. This requires mindful understanding and approval of results in mammalian models to guarantee significance and materialness.

2.2 Hereditary Changeability:

Angelfish populaces show hereditary inconstancy, and the effect of hereditary contrasts on sickness models or medication reactions should be painstakingly thought of. Dissimilar to controlled lab strains, wild angelfish might hold onto hereditary variety that confounds the understanding of results. Tending to this challenge includes coordinating hereditary investigations and utilizing progressed genomic devices to grasp the ramifications of hereditary minor departure from research results.

2.3 Natural Responsiveness:

Angelfish are profoundly delicate to ecological circumstances, and changes in temperature, water quality, or living space design can essentially impact their physiology. This postures difficulties in normalizing trial conditions, as varieties in the climate might prompt variable exploratory results.

Creating powerful exploratory conventions that record for ecological awareness and executing controls becomes vital for solid translational experiences.

2.4 Restricted Comprehension of Angelfish Physiology:

In spite of the headway made in angelfish research, there are holes in how we might interpret their physiology, particularly at the sub-atomic and cell levels. The absence of extensive information upsets the extrapolation of discoveries to human wellbeing. Tending to this challenge includes leading top to bottom examinations on angelfish physiology, including genomics, proteomics, and metabolomics, to unwind the complexities of their natural frameworks.

2.5 Moral Contemplations and Preservation:

As angelfish research extends, moral contemplations connected with the utilization of these creatures become principal. Moral treatment of creatures is a fundamental guideline, and scientists should comply to rigid moral norms. Furthermore, there are concerns connected with the protection of angelfish species in their regular environments. Adjusting logical investigation with moral and preservation contemplations is a basic test in angelfish research.

2.6 Restricted Public Mindfulness:

Notwithstanding the likely translational effect of angelfish research, there is restricted public mindfulness and comprehension of its importance. Conveying the pertinence of angelfish studies to the more extensive public is fundamental for gathering support, cultivating joint efforts, and advancing the coordination of angelfish-inferred information into general wellbeing drives. Overcoming any barrier between mainstream researchers and the general population is a key test.

3. Future Headings

3.1 Coordinating Multi-Omics Approaches:

The coordination of multi-omics approaches, including genomics, metabolomics, and transcriptomics, is a promising course to upgrade how we might interpret angelfish science. By utilizing state of the art advancements, scientists can unwind the atomic complexities of angelfish physiology. This far reaching approach won't just address holes in our ongoing information yet in addition give an establishment to additional exact translational extrapolations.

3.2 High level Hereditary Control Methods:

Progressions in hereditary control strategies, like CRISPR/Cas9 innovation, offer the possibility to refine and control hereditary changes in angelfish.

This permits specialists to make more exact sickness models and study explicit qualities' capabilities more meticulously. Further developed hereditary control procedures add to the age of solid information for translational applications.

3.3 Similar Examinations with Mammalian Models:

To upgrade translational pertinence, directing relative examinations among angelfish and deep rooted mammalian models is fundamental. Using both angelfish and mammalian models, like mice or zebrafish, permits analysts to recognize rationed pathways, approve discoveries, and address species-explicit contrasts. Relative examinations add to a more thorough comprehension of organic cycles.

3.4 Laying out Normalized Trial Conditions:

Defeating the test of natural responsiveness requires the foundation of normalized trial conditions. Analysts should foster conventions that control for natural factors, guaranteeing reproducibility and dependability of results. This includes establishing counterfeit conditions that emulate regular circumstances while taking into account exploratory control, empowering predictable and equivalent information across studies.

3.5 Huge Information and Computational Science:

The time of large information and computational science offers integral assets for breaking down complex datasets. Applying these ways to deal with angelfish research empowers the ID of examples, connections, and prescient models. Incorporating enormous information examination and computational demonstrating upgrades our capacity to extricate significant bits of knowledge from huge scope natural datasets, adding to a more extensive comprehension of angelfish science.

3.6 Cooperative Exploration Drives:

Working with cooperative exploration drives that include multidisciplinary groups is significant for propelling angelfish research. Coordinated efforts between researcher, geneticists, ecological researchers, and clinicians consider an all encompassing investigation of angelfish science and its translational ramifications. These drives advance information trade, expertise sharing, and a synergistic way to deal with tending to complex exploration questions.

3.7 Public Commitment and Schooling:

To address the test of restricted public mindfulness, proactive public commitment and schooling drives are fundamental. Specialists ought to effectively convey their discoveries to the general population, underscoring the possible effect on human wellbeing.

Public commitment cultivates support for angelfish research, energizes capable and moral practices, and improves the incorporation of angelfish-inferred information into more extensive wellbeing conversations.

3.8 Moral Structures and Protection Endeavors:

Creating strong moral systems for angelfish research and executing protection endeavors are basic for mindful logical investigation. Specialists ought to stick to severe moral principles, guaranteeing the sympathetic treatment of angelfish and limiting any likely adverse consequence on wild populaces. Protection drives can include territory safeguarding, capable reproducing practices, and backing for economical exploration rehearses.

4. Applications in Human Wellbeing

4.1 Medication Revelation and Therapeutics:

Angelfish research holds huge potential for drug disclosure and remedial turn of events. By recognizing saved pathways and testing drug applicants in angelfish models, scientists can smooth out the medication improvement process. The translational application includes approving the viability and security of recognized intensifies in mammalian models prior to advancing to human clinical preliminaries.

4.2 Accuracy Medication Approaches:

Bits of knowledge from angelfish research add to the improvement of accuracy medication draws near. Understanding hereditary varieties and their effect on drug reactions takes into account customized remedial mediations. By fitting medicines in light of individual qualities, accuracy medication improves remedial adequacy while limiting unfriendly impacts, lining up with the standards of customized medical services.

4.3 Illness Demonstrating and Grasping Pathogenesis:

Angelfish act as important models for grasping infection pathogenesis. By actuating explicit hereditary alterations or natural irritations, analysts can show human illnesses and analyze the fundamental systems. This information is instrumental in disentangling the intricacies of sicknesses, distinguishing restorative targets, and creating designated mediations for conditions going from metabolic problems to cardiovascular illnesses.

4.4 General Wellbeing Drives:

The information got from angelfish examination can illuminate general wellbeing drives pointed toward forestalling and overseeing infections. Bits of knowledge into the effect of diet, way of life, and natural elements on angelfish digestion add to prove based proposals for human populaces. General wellbeing efforts can use this data to advance better ways of life, diminish sickness hazard, and upgrade by and large prosperity.

5. Moral Contemplations in Translational Exploration

5.1 Moral Treatment of Creatures:

Guaranteeing the moral treatment of angelfish and other lab creatures is an essential standard in translational examination. Specialists should comply with severe moral norms, including sympathetic lodging conditions, proper consideration, and adherence to rules laid out by moral survey sheets. Moral treatment stretches out to limiting uneasiness, decreasing the quantity of creatures utilized, and focusing on their government assistance.

5.2 Informed Assent in Human Examinations:

While interpreting discoveries from angelfish exploration to human clinical examinations, it is fundamental to acquire informed assent. Moral contemplations include straightforwardly imparting the idea of the exploration, likely dangers, and advantages to human members. Informed assent guarantees that people know about the review's goals and enthusiastically partake in research that might impact their wellbeing.

5.3 Natural and Protection Morals:

Translational exploration including angelfish ought to consider natural and protection morals. Capable practices incorporate limiting the effect on wild populaces, supporting preservation endeavors, and carrying out feasible exploration rehearses. Scientists ought to effectively take part in protection drives and add to the safeguarding of biological systems where angelfish species dwell.

8.1Addressing challenges in translating angelfish findings to human health.

Angelfish, as a model organic entity, has given priceless bits of knowledge into different features of science, from hereditary qualities to digestion. The translational capability of these discoveries to human wellbeing is gigantic, promising progressions in figuring out sicknesses, drug revelation, and helpful mediations. In any case, the interpretation of angelfish exploration to significant applications in human wellbeing faces a few difficulties. This investigation digs into the diverse difficulties experienced in deciphering angelfish discoveries and proposes techniques to defeat these obstacles.

2. Species-Explicit Contrasts

2.1 Protection of Essential Cycles:

While angelfish and people share essential natural cycles, species-explicit contrasts present a test in extrapolating discoveries straightforwardly. The preservation of fundamental pathways, like metabolic cycles and hereditary flagging, gives a premise to translational endeavors.

In any case, understanding the subtleties of these distinctions is urgent to guaranteeing the pertinence and materialness of angelfish-determined bits of knowledge to human wellbeing.

2.2 Genomic Inconstancy:

Hereditary fluctuation inside angelfish populaces presents intricacies in making an interpretation of hereditary discoveries to human genomics. Not at all like controlled lab strains, wild angelfish might show hereditary variety that influences illness models and medication reactions. Tending to this challenge includes thinking about hereditary varieties, using progressed genomic apparatuses, and leading near investigations to perceive moderated pathways with importance to human hereditary qualities.

3. Ecological Awareness

3.1 Normalizing Trial Conditions:

Angelfish are exceptionally delicate to natural circumstances, and varieties in elements like temperature, water quality, and territory construction can fundamentally impact exploratory results. Normalizing trial conditions becomes pivotal to guarantee the reproducibility and unwavering quality of results. Creating conventions that copy normal circumstances while giving exploratory control is fundamental to beating the test of natural responsiveness.

3.2 Controlled Research facility Conditions:

Establishing controlled lab conditions that duplicate explicit parts of angelfish territories permits scientists to limit the effect of ecological changeability. This includes careful regard for water quality, temperature, and different elements that impact angelfish physiology. Via cautiously dealing with these factors, scientists can upgrade the dependability of exploratory information and work with additional precise translational extrapolations.

4. Restricted Comprehension of Angelfish Physiology

4.1 Far reaching Omics Approaches:

To address holes in how we might interpret angelfish physiology, taking on complete omics approaches is basic. Genomics, transcriptomics, and metabolomics examinations can give a definite sub-atomic profile of angelfish science. This multi-omics mix improves how we might interpret the mind boggling networks overseeing angelfish physiology and adds to additional precise translational understandings.

4.2 Near Physiology Studies:

Leading near physiology concentrates among angelfish and warm blooded animals helps with clarifying the similitudes and contrasts in physiological cycles.

This approach includes efficiently dissecting organ frameworks, cell capabilities, and atomic pathways in both angelfish and people. Near examinations offer a comprehensive perspective on physiological transformations and illuminate scientists about expected difficulties in interpretation.

5. Moral Contemplations and Preservation

5.1 Moral Treatment of Angelfish:

Keeping up with moral principles in angelfish research is fundamental for capable logical investigation. Analysts should focus on the moral treatment of angelfish, guaranteeing accommodating circumstances, limiting pressure, and complying to laid out moral rules. Moral contemplations reach out to the utilization of hereditary control procedures, where moral structures ought to direct the mindful utilization of these advances.

5.2 Protection of Angelfish Species:

Protection endeavors are essential to tending to moral contemplations and guaranteeing the manageability of angelfish research. Analysts ought to effectively add to preservation drives that emphasis on safeguarding angelfish species in their normal living spaces. Joint effort with ecological associations, adherence to maintainable practices, and promotion for dependable exploration add to the drawn out preservation of angelfish populaces.

6. Procedures for Defeating Translational Difficulties

6.1 Coordinating Multi-Omics Information:

The joining of multi-omics information, including genomics, transcriptomics, and metabolomics, is a strong procedure to conquer the test of restricted comprehension of angelfish physiology. By creating far reaching datasets, specialists can uncover unpredictable atomic communications and distinguish monitored pathways. Incorporating multi-omics information improves the profundity of information and works with additional precise translational understandings.

6.2 Cooperative Exploration Organizations:

Encouraging cooperative examination networks that include specialists from assorted disciplines is vital for tending to translational difficulties. Joint efforts between scholars, geneticists, ecological researchers, clinicians, and ethicists unite corresponding mastery. These organizations work with information trade, upgrade exploratory plan, and add to a more comprehensive way to deal with angelfish research.

6.3 High level Hereditary Control Methods:

Progressions in hereditary control procedures, especially CRISPR/Cas9 innovation, offer chances to refine and control hereditary adjustments in angelfish. Further developed accuracy in hereditary control permits analysts to make more exact sickness models and explore explicit qualities' capabilities more meticulously. This headway adds to the age of solid information for translational applications.

6.4 Near Investigations with Mammalian Models:

To improve translational significance, directing relative examinations among angelfish and laid out mammalian models is fundamental. This approach includes utilizing both angelfish and mammalian models, like mice or zebrafish, to recognize rationed pathways, approve discoveries, and address species-explicit contrasts. Near investigations add to a more complete comprehension of natural cycles.

6.5 Public Commitment and Schooling:

Tending to the test of restricted public mindfulness requires proactive public commitment and schooling drives. Analysts ought to effectively convey the significance of angelfish exploration to human wellbeing, underscoring likely commitments to sickness grasping, drug revelation, and restorative intercessions. Public commitment drives cultivate support for mindful examination practices and increment consciousness of the translational effect of angelfish discoveries.

7. Applications in Human Wellbeing

7.1 Medication Revelation and Therapeutics:

Conquering translational difficulties in angelfish research holds huge potential for drug disclosure and remedial turn of events. Recognizing rationed pathways and testing drug applicants in angelfish models smoothes out the medication improvement process. Translational application includes approving the viability and wellbeing of recognized intensifies in mammalian models prior to advancing to human clinical preliminaries.

7.2 Illness Demonstrating and Restorative Targets:

Angelfish act as important models for sickness demonstrating, helping with the distinguishing proof of restorative targets. By prompting explicit hereditary changes or natural bothers, analysts can demonstrate human infections and analyze the hidden components. This information is instrumental in disentangling the intricacies of illnesses and creating designated mediations for conditions going from metabolic problems to cardiovascular sicknesses.

7.3 General Wellbeing Drives:

The information got from angelfish examination can illuminate general wellbeing drives pointed toward forestalling and overseeing infections. Experiences into the effect of diet, way of life, and natural elements on angelfish digestion add to confirm based proposals for human populaces. General wellbeing efforts can use this data to advance better ways of life, diminish illness chance, and upgrade by and large prosperity.

8. Moral Contemplations in Translational Exploration

8.1 Moral Treatment of Creatures:

Guaranteeing the moral treatment of angelfish and other lab creatures is a central guideline in translational exploration. Analysts should comply with severe moral norms, including altruistic lodging conditions, suitable consideration, and adherence to rules laid out by moral audit sheets. Moral treatment reaches out to limiting inconvenience, lessening the quantity of creatures utilized, and focusing on their government assistance.

8.2 Informed Assent in Human Examinations:

While deciphering discoveries from angelfish examination to human clinical investigations, it is fundamental to get educated assent. Moral contemplations include straightforwardly conveying the idea of the examination, likely dangers, and advantages to human members. Informed assent guarantees that people know about the review's targets and eagerly take part in research that might impact their wellbeing.

8.3 Natural and Preservation Morals:

Translational exploration including angelfish ought to consider natural and preservation morals. Mindful practices incorporate limiting the effect on wild populaces, supporting protection endeavors, and carrying out reasonable examination rehearses. Scientists ought to effectively take part in protection drives and add to the safeguarding of biological systems where angelfish species live.

8.2 Proposing future directions for research and collaboration.

Angelfish research has arisen as a critical field, offering bits of knowledge into major organic cycles with expected applications in human wellbeing. As we stand at the convergence of disclosure and translational applications, significant to imagine future bearings drive angelfish examination into new domains of understanding. This investigation frames key regions for future exploration and proposes methodologies for cultivating cooperative endeavors that rise above disciplinary limits.

2. Disentangling the Wildernesses of Angelfish Physiology

2.1 Top to bottom Omics Studies:

The fate of angelfish research lies in directing top to bottom omics studies to disentangle the complexities of their physiology. Incorporating genomics, transcriptomics, proteomics, and metabolomics will give an extensive sub-atomic profile, revealing insight into the hereditary guideline, cell processes, and metabolic pathways intended for angelfish. This multi-omics approach frames the establishment for understanding how angelfish adjust to their surroundings and the ramifications for human wellbeing.

2.2 Near Physiology Across Species:

Growing relative physiology concentrates past angelfish to incorporate assorted fish species and warm blooded creatures improves how we might interpret transformative variations. By examining similitudes and contrasts in physiological reactions, analysts can recognize rationed pathways and species-explicit subtleties. This relative methodology extends our insight into angelfish as well as adds to a more extensive comprehension of vertebrate physiology.

3. Coordinating High level Hereditary Control Procedures

3.1 Refining CRISPR/Cas9 Innovation:

Headways in hereditary control procedures, especially the CRISPR/Cas9 framework, offer extraordinary open doors for refining hereditary adjustments in angelfish. Future examination ought to zero in on improving the accuracy, effectiveness, and adaptability of hereditary control apparatuses. This incorporates creating procedures for tissue-explicit changes, inducible quality articulation frameworks, and strategies for enormous scope genomic modifications.

3.2 Practical Genomics for Designated Examinations:

The use of utilitarian genomics draws near, for example, CRISPR-based screens, empowers designated examinations concerning the elements of explicit qualities in angelfish. By deliberately bothering qualities and surveying the phenotypic results, scientists can translate the jobs of individual qualities in different natural cycles. This information is significant for propelling comprehension we might interpret quality capability and its suggestions for human wellbeing.

4. Cooperative Exploration Organizations: Building Extensions Across Disciplines

4.1 Multidisciplinary Coordinated efforts:

The eventual fate of angelfish research lies in cultivating multidisciplinary joint efforts that rise above customary limits. Cooperative organizations ought to unite specialists from assorted fields, including science, hereditary qualities, natural science, computational science, and medication. By coordinating mastery from various disciplines, scientists can handle complex inquiries, plan more hearty analyses, and produce comprehensive experiences into angelfish science.

4.2 Collaborations with Human Wellbeing Scientists:

Laying out coordinated efforts with specialists zeroed in on human wellbeing is fundamental for making an interpretation of angelfish discoveries into significant applications. Drawing in with clinicians, pharmacologists, and experts in different clinical fields empowers the ID of expected applications in drug disclosure, illness displaying, and restorative mediations. Overcoming any issues between angelfish examination and human wellbeing requires a purposeful work to work with cross-disciplinary exchange and cooperation.

5. Natural Preservation and Moral Contemplations

5.1 Economical Exploration Practices:

The eventual fate of angelfish research should focus on economical practices that limit the natural effect of logical investigation. Analysts ought to effectively add to ecological protection endeavors, supporting for dependable assortment rehearses, and partaking in territory safeguarding drives. Feasible examination rehearses guarantee the life span of angelfish populaces in their normal living spaces.

5.2 Moral Structures for Hereditary Control:

As hereditary control methods advance, scientists should lay out powerful moral structures to direct their capable use. Future exploration ought to include ethicists and bioethics specialists in creating rules that guarantee the empathetic treatment of angelfish and the moral utilization of hereditary adjustments. Moral contemplations ought to reach out to the expected ramifications of hereditary controls on wild populaces.

6. Information Sharing and Open Science Drives

6.1 Open Access Data sets:

The eventual fate of angelfish research is interwoven with the standards of open science. Laying out open-access information bases that house angelfish genomic, transcriptomic, and phenotypic information advances straightforwardness and works with coordinated effort. By making information broadly available, specialists from around the world can add to the aggregate comprehension of angelfish science and speed up logical advancement.

6.2 Cooperative Information Examination Stages:

Creating cooperative information examination stages improves the aggregate force of angelfish research. Cloud-based stages that permit specialists to share datasets, lead joint examinations, and team up continuously cultivate a feeling of local area and speed up the speed of revelation. These stages ought to focus on information security, protection, and adherence to moral guidelines.

7. Public Commitment and Science Correspondence

7.1 Instructive Effort Projects:

The fate of angelfish research requires dynamic commitment with general society to cultivate understanding and appreciation for the significance of this logical undertaking. Instructive effort programs, including school visits, studios, and public talks, can demystify the universe of angelfish research and rouse the up and coming age of researchers. Connecting with the general population in the energy of logical revelation constructs a strong local area around angelfish research.

7.2 Science Correspondence Drives:

Powerful science correspondence is central for passing the significance and effect of angelfish research on to a more extensive crowd. Future endeavors ought to include devoted science correspondence experts who can make an interpretation of mind boggling logical ideas into open and connecting with content. Using different media, including social stages, webcasts, and narratives, improves the scope and effect of science correspondence drives.

8. Financing Drives and Asset Assignment

8.1 Designated Exploration Financing:

To drive angelfish investigation into what was in store, designated financing drives are fundamental. Subsidizing offices ought to focus on research proposition that exhibit advancement, interdisciplinary joint effort, and translational potential. Putting resources into projects that investigate the utilizations of angelfish research in human wellbeing and ecological protection lines up with more extensive cultural objectives.

8.2 Asset Designation for Long haul Studies:

Long haul reads up are imperative for disentangling the intricacies of angelfish science and grasping the ramifications of hereditary changes. Asset portion ought to focus on supported financing for projects that require consistent observing, like investigations on the impacts of ecological changes and hereditary modifications overstretched periods. Long haul studies add to a more nuanced comprehension of angelfish physiology.

8.3 Ethical considerations and implications for medical practice.

Morals in clinical practice frames the bedrock of patient consideration, research, and the more extensive medical care framework. As medical services experts explore the complicated scene of clinical direction, therapy conventions, and headways in clinical innovation, moral contemplations assume a significant part in guaranteeing the prosperity of patients and maintaining the trustworthiness of the clinical calling. This investigation digs into the moral components of clinical work on, looking at key contemplations and their suggestions for the conveyance of value medical services.

2. Informed Assent: Foundation of Patient Independence

2.1 Definition and Significance:

Informed assent is a key moral rule that supports patient independence. It includes giving patients exhaustive data about their ailment, proposed medicines, expected dangers, and options, empowering them to settle on informed conclusions about their consideration. Regarding patient independence isn't just a moral commitment yet additionally a lawful prerequisite in numerous medical services locales.

2.2 Difficulties and Correspondence Systems:

Challenges in getting legitimate informed assent emerge from variables like language hindrances, mental disabilities, and the intricacy of clinical data. Medical care experts should utilize powerful correspondence methodologies, including the utilization of plain language, visual guides, and translators when vital, to guarantee that patients completely grasp the data introduced. The objective is to engage patients to effectively take part in choices in regards to their wellbeing.

2.3 Shared Navigation:

The development of clinical morals has prompted the advancement of shared decision-production as a cooperative methodology between medical services suppliers and patients. This model perceives the ability of the two players and includes open correspondence, shared regard, and thought of the patient's qualities and inclinations. Shared dynamic encourages a more understanding focused way to deal with clinical consideration, lining up with the standards of helpfulness and non-perniciousness.

3. Value and Non-Wrathfulness: Difficult exercises in Tolerant Consideration

3.1 Advantage:

The moral guideline of advantage stresses the commitment of medical care experts to act to the greatest advantage of their patients. This incorporates giving capable and merciful consideration, ceaselessly refreshing clinical information, and supporting for the prosperity of patients. Value reaches out past the clinical setting, enveloping endeavors to address social determinants of wellbeing and advancing general wellbeing drives.

3.2 Non-Evil:

Non-evil, frequently communicated through the proclamation "cause no damage," requires medical care experts to focus on quiet security and try not to inflict damage, whether through purposeful activities or oversights. This standard aides choices in regards to analytic techniques, medicines, and mediations. Finding some kind of harmony among usefulness and non-wrathfulness is a fragile moral dance, stressing the requirement for proof based rehearses and a pledge to limiting dangers.

3.3 Moral Predicaments and Dynamic Structures:

Moral predicaments might emerge when advantage and non-wrathfulness collide. Choices in regards to life-supporting medicines, end-of-life care, and asset allotment present difficulties that require cautious thought of moral structures. Using apparatuses, for example, the four standards approach (independence, value, non-perniciousness, equity) helps medical care experts in exploring complex moral choices.

4. Equity: Value and Decency in Medical care

4.1 Distributive Equity:

The rule of equity tends to the fair conveyance of medical care assets, guaranteeing that all people have equivalent admittance to essential clinical consideration. Distributive equity turns out to be especially striking in conversations about medical services variations, asset allotment during general wellbeing emergencies, and the improvement of medical care arrangements. Making progress toward value in medical care conveyance lines up with the moral basic of equity.

4.2 Wellbeing Variations and Social Capability:

Addressing wellbeing variations requires medical care experts to take part in social ability, perceiving and regarding the different foundations, convictions, and encounters of patients. Social skill includes fitting medical care ways to deal with meet the extraordinary necessities of people from various ethnic, social, and financial foundations. This obligation to equity includes pushing for arrangements that kill fundamental hindrances to medical services access and quality.

4.3 Asset Assignment and Moral Independent direction:

In circumstances of asset shortage, for example, during a pandemic or in low-asset settings, medical services experts face moral problems in designating restricted assets. Dynamic structures, for example, utilitarianism or the fair innings approach, guide the fair dissemination of assets in light of rules like clinical need, potential for benefit, and the standard of saving the most lives.

5. Protection and Classification: Shielding Patient Data

5.1 Significance of Protection:

Regarding patient protection and keeping up with classification are basic parts of moral clinical practice. Patients depend medical care suppliers with touchy and individual data, and safeguarding this data is fundamental for building trust and protecting the patient-supplier relationship. Protection contemplations reach out to electronic wellbeing records, specialized techniques, and conversations inside medical services groups.

5.2 Difficulties in the Advanced Age:

The advanced change of medical services carries the two advantages and difficulties to patient protection. Electronic wellbeing records, telemedicine stages, and wellbeing data trades upgrade availability yet in addition raise worries about information security and unapproved access. Medical services experts should explore the moral components of innovation, guaranteeing that patient data stays secure and private.

5.3 Informed Assent for Information Use:

As medical care information turns out to be progressively important for exploration and quality improvement drives, getting educated assent for information use is vital. Patients ought to be educated about how their information will be utilized, expected chances, and the shields set up to safeguard their security.

Straightforwardness in information rehearses lines up with the standards of independence and regard for people's privileges.

6. End-of-Life Care and Advance Orders

6.1 Development Orders:

Advance orders, like living wills and sturdy legal authority for medical care, permit people to communicate their inclinations with respect to clinical therapy in case of weakening. Regarding and executing advance orders is a moral commitment for medical services suppliers, guaranteeing that patients' desires are respected, especially in choices about existence supporting therapies and end-of-life care.

6.2 Palliative Consideration and Hospice:

Integrating palliative consideration standards into clinical practice underlines the help of anguish and improvement of personal satisfaction for patients with difficult sicknesses. Hospice care, zeroing in on solace and respect in the last phases of life, mirrors a promise to non-surrender and sympathetic finish of-life care. Moral contemplations include open correspondence with patients and their families, tending to their qualities and inclinations.

7. Arising Advances and Moral Wildernesses

7.1 Genomic Medication:

The coordination of genomic data into clinical practice raises moral contemplations connected with protection, informed assent, and the potential for hereditary segregation. Medical services experts should explore the intricacies of hereditary testing, guaranteeing that patients are very much educated about the ramifications regarding genomic data for them as well as their families.

7.2 Man-made brainpower in Medical care:

The utilization of man-made brainpower (artificial intelligence) in medical services, from demonstrative devices to therapy proposals, acquaints moral contemplations related with straightforwardness, responsibility, and predisposition. Medical services experts taking on simulated intelligence advances should consider the moral ramifications of depending on algorithmic independent direction, resolving issues of decency and patient trust.

8. Moral Schooling and Consistent Expert Turn of events

8.1 Joining of Morals in Clinical Training:

Guaranteeing that medical services experts are knowledgeable in moral standards requires the coordination of morals schooling all through clinical preparation. Clinical schools and residency projects ought to underscore the significance of moral navigation, relational abilities, and social capability. Moral difficulties, case-based conversations, and associations with assorted patient populaces add to moral capability.

8.2 Consistent Expert Improvement in Morals:

The powerful idea of medical services, combined with advancing moral difficulties, highlights the significance of consistent expert improvement in morals.

Medical services suppliers ought to participate in continuous schooling, partake in moral case conversations, and remain informed about arising moral issues in medication. Proficient associations and establishments assume an essential part in giving assets and backing to moral turn of events.